Critical Challenges in Law and Government: Re-enacting Confederation—A Simulation

This series is one of a number of initiatives organized under the auspices of *TC*² — *The Critical Thinking Cooperative*. *TC*² is a partnership of British Columbia school districts, faculties of education, teacher professional associations, and other educational organizations. The aim of the cooperative is to promote critical thinking through partner-sponsored professional development, resource development and research activities. *TC*² functions as a clearinghouse and support network to promote critical thinking from primary to post-secondary education.

For more information about *TC*² contact:

LeRoi Daniels
The Critical Thinking Cooperative
c/o Richmond School District
7811 Granville Avenue
Richmond, British Columbia V6Y 3E3
Phone: (604) 732-1907 Fax: (604) 732-1957
e-mail: leroi.daniels@ubc.ca

Critical Challenges Across the Curriculum Series

Critical Challenges in Law and Government: Re-enacting Confederation—A Simulation

Greg Buium
Janet Thompson

with contributions from

Roland Case
Tom Morton
Phyllis Schwartz

The Critical Thinking Cooperative

Field Programs
Faculty of Education
Simon Fraser University

Series published by

The Critical Thinking Cooperative
c/o Richmond School District
7811 Granville Avenue
Richmond, British Columbia V6Y 3E3

Series distributed by

Pacific Educational Press
Faculty of Education
University of British Columbia
Vancouver, British Columbia, V6T 1Z4
Telephone: (604) 822-5385
Facsimile: (604) 822-6603
e-mail: cedwards@interchange.ubc.ca

Series produced by

Field Programs
Faculty of Education
Simon Fraser University
Burnaby, British Columbia, V5A 1S6

Design and Production: M. Kathie Wraight
Project Coordination: Patricia Holborn

Cover Design: Dynamic Desktop

Developed with the support of the Centre for Education, Law and Society (Simon Fraser University)

Reprinted 1999 with minor revisions.

Canadian Cataloguing in Publication Data

Buium, Greg, 1966-
Critical challenges in law and government: re-enacting Confederation

(Critical challenges across the curriculum series, ISSN 1205-9730 ; v. 4A)
Co-published by: Field Relations and Teacher In-Service Education, Faculty of Education, Simon Fraser University.
ISBN 0-86491-194-7

1. Canada—Confederation, 1867—Study and teaching (Secondary) I. Thompson, Janet, 1970- II. Case, Roland, 1951- III. Morton, Tom, 1948- IV. Critical Thinking Cooperative. V. Simon Fraser University. Faculty of Education. Field Relations and Teacher In-Service Education. VI. Title VII. Series.
FC470.B84 1998 971.04'9 C98-910958-5
F1032.B84 1998

Contents

Foreword

Critical Challenges Across the Curriculum is an ongoing series of teacher resources focussed on infusing critical thinking into every school subject. Two features distinguish this series from many other publications on critical thinking—our *curriculum embedded* approach and our emphasis on *teaching the intellectual tools.*

Our approach is to embed critical thinking by presenting focussed questions or tasks that invite critical student reflection about the content of the curriculum. We do not support the view of critical thinking as a set of generic skills or processes that can be developed independent of content and context. Nor do we believe that critical thinking can adequately be addressed as an add-on to the curriculum. Rather, if it is to take a central place in the classroom, critical thinking must be seen as a way of teaching the content of the curriculum. Teachers can help students understand the subject matter, as opposed to merely recall it, by providing continuing opportunities for thoughtful analysis of issues or problems that are central to the subject matter.

The second distinguishing feature of this series is our emphasis on systematically teaching the intellectual tools for critical thinking. Much of the frustration teachers experience when attempting to engage students in thinking critically stems from students' lack of the required concepts, attitudes, knowledge, criteria or strategies—in short, they lack the tools needed to do a competent job. It is often assumed that the mere provision of invitations to think will improve students' reflective competence. We believe that constructing a thoughtful response is like building a house, in that it is impossible to do a competent job in either case unless one has the necessary tools. For this reason, each critical thinking challenge in the series includes a list of the tools needed to respond competently and, more importantly, activities suggesting how these tools may be taught.

We hope that teachers will find these resources of use in increasing and improving the teaching of critical thinking in their subject areas.

Roland Case & LeRoi Daniels
Series Editors

Acknowledgments

A number of people were central to the completion of this project. We were flattered by the invitation to join TC^2's ongoing *Critical Thinking Across the Curriculum* series, and the encouragement and support we received made the process a real pleasure. The discussion groups Roland Case led were fundamental to our final product. Aric Hayes, Tom Morton and Phyllis Schwartz's input during these informal meetings was invaluable. It was Phyllis who proposed our partnership with TC^2. Peter Seixas was generous with his time; we were mindful of his suggestions on integrating primary sources in the Social Studies classroom. We are grateful to British Columbia teachers Susan Hargraves, Holly Burrows and Melanie Mobbs for the ideas they contributed to the Tool Enhancement Supplement and to Elgin Wyatt of North Battleford, Saskatchewan for his ideas on adopting a Father of Confederation persona. We thank Dan Starling for his drawings of the maps of Canada in the 1860s. As well we appreciate the support of the Vancouver School Board, specifically Valerie Overgaard and Frank McCormick. Finally, we would like to thank our department heads at Lord Byng, Phyllis Schwartz and Andy Johnston, for their support during our hectic first year teaching.

Greg Buium & Janet Thompson

Introduction to Critical Thinking

Understanding Critical Thinking

There are many conceptions or accounts of what is involved in thinking critically. The model described here is more fully explained in two articles in the *Journal of Curriculum Studies* entitled "Common Misconceptions of Critical Thinking" and "A Conception of Critical Thinking" by Sharon Bailin, Roland Case, Jerrold Coombs and LeRoi Daniels. For reasons explained below, we propose the following definition:

> *Critical thinking involves thinking through problematic situations about what to believe or how to act where the thinker makes reasoned judgments that embody the qualities of a competent thinker.*

A person is attempting to think critically when she thoughtfully seeks to judge what would be sensible or reasonable to believe or do in a given situation. The need to reach reasoned judgments may arise in countless kinds of problematic situations such as trying to understand a passage in a text, trying to improve an artistic performance, making effective use of a piece of equipment, or deciding how to act in a delicate social situation. What makes these situations problematic is that there is some doubt as to the most appropriate option.

Critical thinking is sometimes contrasted with problem solving, decision making, issue analysis and inquiry. We see these latter terms for rational deliberation as occasions for critical thinking. In all these situations, we need to think critically about the options. There is limited value in reaching solutions or making choices that are not sensible or reasonable. Thus, the term critical thinking draws attention to the quality of thinking required to competently pose and solve problems, reach sound decisions, identify and resolve issues, plan and conduct thoughtful inquiries and so on. In other words, thinking critically is a way of carrying out these thinking tasks just as being careful is a way of walking down the stairs. Thinking critically is not a unique *type* of thinking that is different from other types of thinking, rather it refers to the *quality* of thinking. The association of critical thinking with being negative or judgmental is misleading, since the reference to critical is to distinguish it from uncritical thinking—thinking that accepts conclusions at face value without any assessment of their merits or bases. It is more fruitful to interpret critical in the sense of critique—looking at the merits and shortcomings of alternatives in order to arrive at a reasoned judgment.

Our focus on the quality of thinking does not imply that students must arrive at a preconceived right answer, rather we look to see that their responses exhibit the qualities that characterize good thinking in a given situation. For example, students' responses to a request to cooperatively plan a field trip through group discussion may be judged in light of the accuracy and adequacy of information, how seriously students considered the ideas of others, willingness to express their own ideas, and respect for the ideas of those with whom they disagree. These are all qualities that a competent thinker would exhibit. Similarly, a critically thoughtful response to a newspaper editorial would likely include the following characteristics: sensitivity to any bias on the part of the writer, adequate consideration of alternative points of view, attention to the clarity of definition of key concepts, and assessment of evidence offered in support of the writer's position. We believe that emphasis on qualities that student responses should exhibit focuses teachers' attention on the crucial dimension in promoting and assessing students' competence in thinking critically. The challenge for teachers is to adopt practices that will effectively promote these qualities in their students.

Promoting Critical Thinking

To help students improve as critical thinkers, we propose a four-pronged approach:

- build a *community of thinkers* within the school and classroom;
- infuse opportunities for critical thinking—what we call *critical challenges*—throughout the curriculum;
- develop the *intellectual tools* that will enable students to become competent critical thinkers;
- on a continuing basis *assess students' competence* in using the intellectual tools to think through critical challenges.

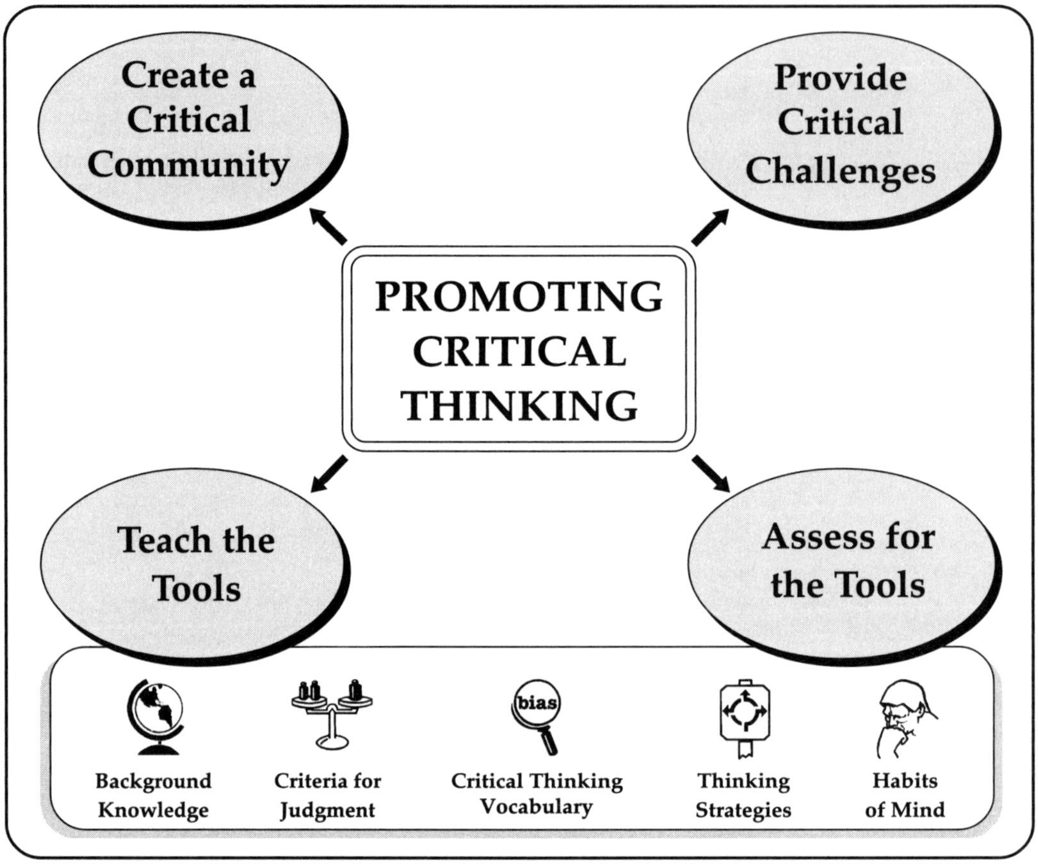

Building a community of thinkers

It is essential to infuse the expectations and opportunities to think critically throughout students' school lives. Developing supportive school and classroom communities where reflective inquiry is valued may be the most important factor in nurturing critical thinking. Many of the tools of critical thinking will not be mastered by students unless their use is reinforced on an ongoing basis. As well, the image of the thinker as a solitary figure is misleading. No one person can perfectly embody all the desired attributes—we must learn to rely on others to complement our own thoughts. Students will not learn how to be contributors to, and beneficiaries of, a community of thinkers unless they become members in that kind of community. There are many routines and norms that teachers can adopt to create this sense of community:

- As a matter of course, questions and assignments should have built-in expectations to think through, and not merely recall, what is being learned.
- Ongoing opportunities to engage in critical and cooperative dialogue—to confer, inquire, debate and critique—are key to creating a community of thinkers.

- Self- and peer-evaluation are excellent ways of involving students in thinking critically about their own work.
- Teacher modelling of good critical thinking practices is important. Students are more likely to learn to act in desired ways if they see teachers making every effort to be open-minded, to seek clarification where needed, to avoid reaching conclusions based on inadequate evidence, and so on. Opportunities to see their teachers as co-investigators, and not solely as resident experts, may send important messages to students about the need to think critically.

Infusing critical challenges throughout the curriculum

Critical thinking is always in response to a problematic situation, where judgment among alternatives is required. If students are to improve in their ability to think critically, they must have numerous opportunities to engage and think through problematic situations. We label as *critical challenges* those problematic situations which are deliberately presented to students for consideration.

Four questions should guide the choice of critical challenges.

- *Does the question or task require judgment?* Critical thinking occurs only in the context of a problematic situation. If an answer is simply there, waiting to be found, or if any and all answers are acceptable then there is no invitation to think critically. A question or task is a critical challenge only if it invites students to assess the reasonableness of plausible options or alternative conclusions; in short, the task must require more than retrieval of information, rote application of a strategy, uninformed guessing or mere assertion of a preference.

- *Will the challenge be meaningful to students?* We enhance our efforts to promote critical thinking if students find the questions interesting. Trivial, decontextualized mental exercises often alienate or bore students. It is important to frame challenges that are likely to engage students in tackling critical questions and tasks that they will find meaningful.

- *Does the challenge address key aspects of the subject matter?* If we are to make time for critical thinking the focus must be on matters that are at the heart of the curriculum—and not peripheral to it. As well, students are more likely to learn the content of the curriculum if they are invited to think critically about issues embedded in the subject matter.

- *Do students have the tools or can they reasonably acquire the tools needed to competently address the challenge?* Students are more likely to succeed if they possess the tools to deal with the challenge before them. Provision must be made to support students in acquiring the essential tools needed to competently meet the critical challenge. For this reason, challenges should be sufficiently focussed so that students are not overwhelmed by the enormity or complexity of the task.

Developing intellectual tools for thinking critically

The key to helping students develop as critical thinkers is to nurture competent use of five types of tools of thinking. These categories of tools are *background knowledge, criteria for judgment, critical thinking vocabulary, thinking strategies* and *habits of mind.*

Background Knowledge

—the information about a topic required for thoughtful reflection

Students cannot think deeply about a topic if they know little about it. Two questions to ask in developing this tool:

- What background information do students need for them to make a well-informed judgment on the matter before them?
- How can students be assisted in acquiring this information in a meaningful matter?

Criteria for Judgment

—the considerations or grounds for deciding which of the alternatives is the most sensible or appropriate

Critical thinking is essentially a matter of judging which alternative is sensible or reasonable. All judgments are based on criteria of some sort or other. Students need help in thinking carefully about the criteria to use when judging various alternatives.

- Is my estimate *accurate*?
- Is the interpretation *plausible?*
- Is the conclusion *fair* to all?
- Is my proposal *feasible*?

Critical Thinking Vocabulary

—the range of concepts and distinctions that are helpful when thinking critically

Students require the vocabulary or concepts that permit them to make important distinctions among the different issues and thinking tasks facing them. These include the following:

- inference and direct observation;
- generalization and over generalization;
- premise and conclusion;
- bias and point of view.

Thinking Strategies

—the repertoire of heuristics, organizing devices, models and algorithms that may be useful when thinking through a critical thinking problem

Although critical thinking is never simply a matter of following certain procedures or steps, numerous strategies are useful for guiding one's performance when thinking critically:

- *Making decisions:* Are there models or procedures to guide students through the factors they should consider (e.g., a framework for issue analysis or problem solving)?
- *Organizing information:* Would a graphic organizer (e.g., webbing diagrams, Venn diagrams, "pro and con" charts) be useful in representing what a student knows about the issue?
- *Role taking:* Before deciding on an action that affects others, should students put themselves in the others' positions and imagine their feelings?

Habits of Mind

—the values and attitudes of a careful and conscientious thinker

Being able to apply criteria and use strategies is of little value unless students also have the habits of mind of a thoughtful person. These include:

- *Open-minded:* Are students willing to consider evidence opposing their view and to revise their view should the evidence warrant it?
- *Fair-minded:* Are students willing to give impartial consideration to alternative points of view and not simply impose their preference?
- *Independent-minded:* Are students willing to stand up for their firmly held beliefs?
- *Inquiring or "critical" attitude:* Are students inclined to question the clarity of and support for claims and to seek justified beliefs and values?

Assessing for the tools

Assessment is an important complement to the teaching of the tools of critical thinking. As suggested by the familiar adages "What is counted counts" and "Testing drives the curriculum," evaluation has important implications for what students consider important and ultimately what students learn. Evaluations that focus exclusively on recall of information or never consider habits of mind fail to assess, and possibly discourage, student growth in critical reflection.

A key challenge in assessing critical thinking is deciding what to look for in a student's answer. If there is no single correct response, we may well ask: "On what basis, then, can we reliably assess students?" The qualities that we would expect to see exhibited in a successful response should provide the basis for our assessments. In the case of critical thinking, this means we would want to see whether or not students exhibited the qualities of a competent thinker. Thus, the intellectual resources or tools for critical thinking become the criteria for assessing students' work. More specifically, as students respond (orally, in writing, visually, etc.) to a critical challenge we should look for indications of the extent to which their comments or products reveal appropriate use of the desired tools. In other words, has the student:

- provided adequate and accurate information?
- satisfied relevant criteria for judgment?
- revealed understanding of important vocabulary?
- made effective use of appropriate thinking strategies?
- demonstrated the desired habits of mind?

The following example suggests each of the five types of critical thinking tools and specific assessment criteria that might be used in assessing evidence of critical thinking in an argumentative essay and an artistic work.

Type of criteria for assessment	Evidence of critical thinking in a persuasive essay	Evidence of critical thinking in an artistic work
Background Knowledge	• cited accurate information.	• revealed knowledge of the mechanics of the medium.
Criteria for Judgment	• provided ample evidence; • arranged arguments in logical sequence.	• work was imaginative; • work was clear and forceful.
Critical Thinking Vocabulary	• correctly distinguished "arguments" from "counter arguments."	• represented "point of view."
Thinking Strategies	• used appropriate strategies for persuasive writing.	• employed suitable rehearsal/ preparation strategies.
Habits of Mind	• demonstrated an openness to alternative perspectives; • refrained from forming firm opinions where the evidence was inconclusive.	• was open to constructive criticism; • demonstrated a commitment to high quality; • demonstrated a willingness to take risks with the medium.

Peer- and self-assessment are especially effective means of encouraging critical thinking since the assessment of an assignment is a synonym for thinking critically about the assignment. The box below contains statements about several thinking strategies and habits of mind that students might consider as they self-assess their thoughtfulness as independent learners.

Self-assessment questions	Implied criteria for assessment
I thought about what I was expected to do before I started.	*Thinking Strategy:* before beginning, get clear about expectations
I asked for help whenever I needed it.	*Thinking Strategy:* ask for help
I stayed "on-task" until all my work was finished.	*Habit of Mind:* intellectual work ethic
I tried my best.	*Habit of Mind:* intellectual work ethic

Overview of Simulation

Critical Challenge

As representatives of a British North American colony at a hypothetical 1864 Ottawa Conference, students decide whether or not to join a proposed Canadian Confederation under the conditions outlined in a draft British North America Act which they have helped to negotiate.

Synopsis

This 8 - 13 hour simulation can be used as the cornerstone for an entire unit of study on Confederation. Students are assigned to one of six groups, each representing a British North American colony at the time of Confederation. They simulate the negotiations that led to Confederation by participating in a fictitious Ottawa Conference held in 1864. The simulation unfolds in five stages as summarized in the following chart

Orientation (1.5 - 2.5 hours)	In the introductory lesson students begin to think about the terms of Confederation by considering current difficulties facing Canada and by discussing the idea of re-visiting the original terms of the union of Canada. Students are introduced to the main tasks of the simulation and assigned to represent one of the six colonies that will take part in the conference.
Preliminary Proposals (3 - 6 hours)	Based on supplied briefing sheets and independent library research, each delegation composes a preliminary list of conditions, with accompanying justifications, for joining the proposed federal union. Students explain and defend their preliminary proposal in a written submission and a five-minute rehearsed presentation. Other delegations ask the presenting colony one question about a pertinent issue.
Final Proposals (2 - 3 hours)	Each delegation discusses the preliminary proposals from the other colonies. Individual representatives become experts on one other colony and then brief fellow delegates on that colony's needs and issues. Students discuss and re-negotiate their preliminary requirements in order to write a final proposal that would appeal to the other colonies but still serve their colony's essential interests. Each delegation prepares and delivers a five-minute presentation on its final proposal. The colonies then respond to each of the other colonies' proposals.
Voting on the Bill (1 hour)	The teacher prepares and presents to the colonies a Confederation Bill based on the final proposals from all the colonies. This bill outlines the terms and conditions of a proposed federal union. Each delegation deliberates and must eventually decide whether or not to accept the terms of the Canadian Confederation. Students are encouraged to reach a consensus decision within their delegations. Those colonies deciding to join Confederation sign the re-negotiated British North America Act, 1867.
Debriefing (1 hour)	Students are encouraged in the final debriefing stage to reflect on their experiences as participants in the exercise and on lessons learned about negotiating agreements. They also undertake a peer- and self-assessment of individual contributions to their group during the simulation.

Tool Enhancement Supplement

During the simulation students will encounter a number of tasks, each requiring a different array of tools. Some required tools, such as knowledge of the colony and its concerns, are built into the basic structure of the simulation. Additional tools are described in the last section of the book. This section, titled *Tool Enhancement Supplement,* offers suggestions for introducing students to a range of tools for four tasks embedded in this simulation:

- contributing to the group;
- developing arguments;
- presenting ideas;
- negotiating agreements.

Tool Enhancement

The *Tool Enhancement* icon (illustrated on the right) indicates where in the course of the simulation the tools for the specified task might be introduced. Since it is unrealistic to expect that all the tools can be addressed, select those tasks and tools which will be most beneficial for students, without attempting so much that it interrupts the flow of the simulation. Alternatively, anticipate the need to develop particular tools by assigning some of the *Tool Enhancement* activities prior to introducing the simulation.

Optional Activities

Occasionally, activities are suggested which add to the simulation but are not necessary to its completion. The term *OPTIONAL* is used to indicate where these activities occur.

Assessment

Assessment

Because of the nature of the simulation many of the products—both written and oral—are group developed. Consequently the three main assessment instruments included in the simulation are for group evaluations. A final peer- and self-assessment instrument is effective in accounting for differences in students' individual contributions to their groups. The *Assessment* icon (illustrated on the right) indicates where each of the suggested assessments may be most appropriately employed. The assessment instruments, themselves, are found in the *Blackline Masters* section.

Blackline Masters

This simulation package is self-contained—all the required materials are supplied and ready for duplication in the form of four types of *Blackline Masters*:

- *briefing sheets*: background information on the colonies, key issues, instructions and maps;
- *student activities*: activity sheets and graphic organizers to assist students in completing and recording answers to assigned tasks;
- *support materials:* documents useful in managing the simulation and other artifacts for teacher and/or student use;
- *assessment*: evaluation questions and scoring rubrics for use in assessing various aspects of student achievement.

Teacher Directions

Orientation

Introducing students to the simulation (45 - 60 minutes)

- Introduce the idea of a Confederation simulation by inviting students to brainstorm the political, social, economic and military problems currently facing Canada. In a class discussion, ask students to speculate on which of these problems have endured for over 100 years. Explain that many people today blame some of our on-going problems on the Confederation agreement that was signed by the Fathers of Confederation in 1867. Suggest to students that they have an opportunity to recreate a problem-free Canada by drafting a Confederation bill as representatives of a British North American colony in a fictitious Ottawa Conference held in 1864.

- Assign students to read *Briefing Sheet: The Ottawa Conference* (Blackline Master #1) which outlines the context and main tasks involved in this simulation. Discuss the contents of the Briefing Sheet with the class, answering any questions that students may have about expectations. If the suggested breakdown for student evaluation outlined on the second page of this Briefing Sheet is not appropriate, do not distribute that page.

- To prepare students for this simulation, assign students to read about (1) the circumstances that contributed to the desire for a federal union: political deadlock in the Canadas, the need for a railway and new trade patterns, the threat of American expansion and the changing attitude of Great Britain towards British North America, and (2) the decisions made at the Charlottetown, Quebec and London Conferences that led to Confederation. This background may be found in most standard textbooks dealing with this era.

Organizing colonial delegations (45 - 90 minutes)

- Divide the class into six equal groups that are reasonably balanced in abilities. Each group will represent a British North American colony at the Confederation Conference. (For the purposes of this simulation, the Canadas are divided into two colonies, despite the Act of Union, 1841):
    - Canada East
    - Canada West
    - New Brunswick
    - Newfoundland
    - Nova Scotia
    - Prince Edward Island

- Assemble an appropriate number of Conference folders. The *Support Material: Conference materials* (Blackline Master #23) can be reproduced and pasted on the front of a file folder to create a cover page. It may be preferable to include in this folder all of the documents for the simulation (except for the Assessment Instruments). Alternatively, if this will be daunting to students, distribute the documents only as needed.

  The following documents will be needed by all representatives (one in each folder):

  - one Support Material (Blackline Master #22) which should be the first page of the Conference folder;
  - one Briefing Sheet (Blackline Master #14); students will already have Blackline Master #1;
  - two Student Activities (Blackline Masters #17 and 21);
  - other Student Activities (Blackline Masters #18-20) are group-based and best distributed as needed.

  The following documents vary with the colony:

  - one Background and one Viewpoints Briefing Sheet selected from Blackline Masters #2-13 corresponding to the assigned colony.

  The following documents are optional:

  - *Annotated bibliography on Confederation* (Blackline Master #15); this Briefing Sheet may be used as a teacher resource or distributed to students (in an enriched class) as the suggested booklist for their research;
  - *Father of Confederation profile* (Blackline Master #16); this optional Student Activity is described below.

- OPTIONAL: To enhance the role play aspect of this simulation, assign each student a Father of Confederation persona. Duplicate on card stock the name tags for each persona found in *Support Material: Representatives' name tags* (Blackline Master #24) attaching each with a paper clip to the front of a Conference folder. The 36 representatives of the British North American colonies who attended one or more of the conferences at which the terms of Confederation were agreed upon are listed in the chart on the following page, with the conference(s) that each attended:

  "C" stands for the Charlottetown Conference (September 1864);

  "Q" for the Quebec Conference (October 1864);

  "L" for the London Conference (December 1866).

Fathers of Confederation

Canada East
George-Etienne Cartier (CQL)
Jean-Charles Chapais (Q)
Alexander T. Galt (CQL)
Hector L. Langevin (CQL)
Thomas D'Arcy McGee (CQ)
Sir Etienne-Paschal Taché (Q)

Canada West
George Brown (CQ)
Alexander Campbell (CQ)
James Cockburn (Q)
William P. Howland (L)
John A. Macdonald (CQL)
William Macdougall (CQL)
Oliver Mowat (Q)

New Brunswick
Edward B. Chandler (CQ)
Charles Fisher (QL)
John Hamilton Gray (CQ)
John M. Johnson (CQL)
Peter Mitchell (QL)*
William H. Steeves (CQ)
Samuel Leonard Tilley (CQL)
R.D. Wilmot (L)*

Newfoundland
Frederick B.T. Carter (Q)
Ambrose Shea (Q)

Nova Scotia
Adams G. Archibald (CQL)
Robert B. Dickey (CQ)
William A. Henry (CQL)
Jonathon McCully (CQL)
John W. Ritchie (L)
Charles Tupper (CQL)

Prince Edward Island
George H. Coles (CQ)
John Hamilton Gray (CQ)
Thomas Heath Haviland (Q)
Andrew A. Macdonald (CQ)
Edward Palmer (CQ)
William H. Pope (CQ)
Edward Whelan (Q)*

Although the simulation will require between four and six members for each colony, historically only two observers from Newfoundland attended any of the three conferences. Consequently, "re-assign" up to three of the representatives from New Brunswick and Prince Edward Island (indicated with an asterisk [*]) to Newfoundland so as to provide that colony with a full complement of representatives.

- OPTIONAL: For added realism, ask students to research their persona, answering the questions found in *Student Activity: Father of Confederation profile* (Blackline Master #16). Information about these men can be found in many high school library reference books, including those references described in *Support Material: Annotated bibliography on Confederation* (Blackline Master #15). In some cases, students may be able obtain a picture of their persona.

- Distribute a Conference folder to each student and allow them a few minutes to look over the materials. Then, using an overheard transparency of *Support Material: Conference agenda* (Blackline Master #22) as a guide, briefly preview the activities in the simulation. Emphasize that each colony's proposals should be historically plausible and that decisions must be realistic. Explain the three functions that the teacher plays during the simulation: (1) re-focussing students during the simulation when historical accuracy is put aside, (2) inventing solutions when students have reached an impasse in negotiation and (3) functioning as Queen Victoria's representative in Canada.

Tool Enhancement: contributing to the group

- Direct students to join with the other representatives from their colony. Invite each delegation to select a convenor, or discussion facilitator, encouraging them to discuss the duties they wish the convenor to play and basic procedures they may wish to put in place for facilitating discussions (e.g., everyone gets an equal opportunity to speak, no put-downs). Consider introducing ideas in the Tool Enhancement Supplement for helping students learn to contribute to the group, specifically Activity A: Imagining Success. Throughout the simulation, encourage students to assign students to perform various group management functions (e.g., recorder, spokesperson) as needed and, each day, to alternate the persons performing the functions. Explain to students that they will be assessed at the end of the simulation on their contributions both to the harmonious operations of their group and to the successful completion of the assigned group projects.

Tool Enhancement: contributing to the group

- OPTIONAL: Encourage each delegation to create a distinct space and identity for its group by, among other things, making a sign and, perhaps, a wall poster. If creating a wall poster is a formal assignment, with specific assigned criteria (i.e., the groups must use symbols to clearly and imaginatively represent both the different delegates' interests and the common ties and concerns of the colony). Consider introducing some of the tools related to contributing to the group which are identified in the Tool Enhancement Supplement. The role play described in Activity G: The Secret Mission would be especially relevant in this situation.

Preliminary Proposals

- The first formal round of the Ottawa Conference has four parts: (1) briefing on colonial issues, (2) independent research, (3) preparation of preliminary demands and (4) presentation of preliminary proposal. Referring to the earlier mentioned *Support Material: Conference agenda* (Blackline Master #22) outline for the class the four tasks for Round One.

Briefing on colonial issues (45 - 90 minutes)

- Assign students to read the *Background* and *Viewpoints Briefing Sheets* (selected from Blackline Masters #2-13) corresponding to the colony which they represent and to make notes on its economic activity, key groups and concerns. Employ a Think-Pair Share strategy to facilitate representatives' learning about their assigned colony. Begin by asking students individually to complete in note form *Student Activity: Colonial perspective* (Blackline Master #17). Students then share their notes with another representative from their colony, adding ideas to their Student Activity sheet. This pair (or threesome, in the case of uneven groups) shares its notes with another pair of fellow delegates, in order to produce even more complete sets of notes. (There may be a need for an additional round, where a foursome shares with another foursome.) If the Think-Pair-Share strategy is new to students, after completing the task invite them to discuss their reactions to it, and suggest that the convenor add it to the list of strategies to be used in facilitating future group discussions. Explain to students that later they will be assessed on their knowledge of their colony (and of the other colonies).

Conducting independent research (45 - 120 minutes)

◆ Allow students one or two class periods to conduct independent research to find more specific reasons for and against Confederation for their colony. The references described in *Support Material: Annotated bibliography on Confederation* (Blackline Master #15) are especially appropriate sources. Encourage students to record ideas from their research on their Colonial Perspective activity sheet.

Preparing preliminary demands (45 - 90 minutes)

◆ Ask each delegation to use the ideas found on their Colonial Perspective sheets to collectively brainstorm a master list of conditions for entering Confederation. Encourage delegates to think of all the ways that their colony's needs/concerns could be addressed by joining with other British North American colonies.

◆ On the basis of this master list each delegation is to collectively compile a preliminary set of initial conditions for its entry into Confederation. Distribute to each delegation two copies (for a working and a polished draft) of *Student Activity: Preliminary proposal* (Blackline Master #18). Assign students to record a carefully worded, prioritized set of demands on this sheet, and to develop justifications for each of their demands. In thinking about their demands, delegates may find it helpful to consult the maps found in *Briefing Sheet: Maps of the colonies in 1864* (Blackline Master #14). Each colony is to duplicate copies of its polished Preliminary Proposal for distribution to the other colonies.

In selecting and framing their conditions for joining confederation, stress upon delegates that the following criteria should be:

- *historically plausible:* The conditions should be realistic; although they need not have actually been made at the Charlottetown and Quebec Conferences.

- *thoughtfully addressed to colonial needs:* Students should think carefully about establishing requirements which best serve their colony's needs. These solutions may come in many forms; for example, trade of commodities or resources with other colonies, a federal tax, specific laws or a combination of such solutions. Encourage delegates to think of alternative ways of addressing their colony's needs.
- *clearly and specifically stated:* Students should avoid vague or very general statements.

Suggest that delegations prioritize their conditions into two categories:

- *High priority:* the bottom line demands which must be part of the final package before the delegation approves the entire package;
- *Low priority:* the desired demands which may be traded to achieve the delegation's bottom line requirements.

Explain to students that the preliminary proposal should comprise both the bottom line demands and negotiable demands.

◆ Throughout the Preliminary Proposal stage students will likely want to discuss their concerns and demands with representatives of other colonies. This student initiative is what makes the Ottawa Conference engaging. This interaction should be allowed to occur, but the teacher must assure that discussions are valuable and that time is not wasted arguing for its own sake. It may be useful to review with the class ActivityJ: Three-tiered Bargaining and other strategies for negotiating agreement found in the Tool Enhancement Supplement.

Tool Enhancement: negotiating agreement

◆ Assign each delegation to plan and rehearse a five-minute presentation which outlines its preliminary demands and their justification. Remind students that all members of each delegation must participate meaningfully in the presentation, that the presentation should make effective use of visual aids, that it should be clearly structured and that delivery should be audible and clear. Encourage each delegation to critique its oral presentation prior to actual delivery to the Conference as suggested in Activity I: Productive Peer Critique found in the Tool Enhancement Supplement.

Tool Enhancement: presenting ideas

◆ Each colony will be allowed to ask the presenting colony one question. Delegations must prepare this question in advance. Remind students to be shrewd and ask questions that will help their delegation better construct its final proposals. Ideas described in the Tool Enhancement Supplement, Activity E: Refining Arguments and Counter-arguments, may be valuable here.

Tool Enhancement: developing arguments

Presenting preliminary proposals (45 - 60 minutes)

- Arrange for each delegation to make its five-minute Preliminary Proposal presentation. Questions after each presentation may take an additional fifteen minutes.

- Evaluate the presentation and the contents of the Preliminary Proposal using the rubrics found in *Assessment Instrument: Preliminary proposals* (Blackline Master #27).

preliminary proposals

Final Proposals

- The Final Proposals round of the Ottawa Conference has three parts: (1) study of other colonies' proposals, (2) preparation of final demands and (3) presentation of the final proposal. Referring to the earlier mentioned *Support Material: Conference agenda* (Blackline Master #22), outline for the class the three tasks for this round.

Studying other colonies' proposals (45 - 60 minutes)

- The Final Proposals stage begins with a homework assignment on the other colonies. On the day of the Preliminary Proposal presentations, distribute to each delegation one set of all of the Background and Viewpoints Briefing Sheets and the Preliminary Proposals for the other colonies.Students in each group are to adopt the role of an expert on one of these colonies and read for next class the documents for that colony. Distribute a second copy of *Student Activity: Knowledge of Colonial perspective* (Blackline Master #17) for experts to use in recording their findings. During the following class, direct students to take turns briefing their colonies on their findings.

- OPTIONAL: Evaluate students' knowledge of the Confederation-related concerns and benefits both for their own colony and for the other colonies using *Assessment: Knowledge of Colonial perspective* (Blackline Master #26).

knowledge of colonies

- OPTIONAL: Invite colonies to send individual delegates as negotiators to other groups to learn more about desired or essential modifications, to persuade them to change or to negotiate compromises on their conditions. As the negotiations continue, each delegation is to re-examine conditions and counter by proposing qualifying clauses, compromises and trade-offs. Delegates should consult with other colonies to better understand what may be needed to gain acceptance of key requirements. Negotiating modifications that are acceptable to all parties may be facilitated if students keep in mind the difference between disputes about underlying values/goals and disputes over the strategies to achieve them. (There may be several ways to achieve the goal.)

◆ It may be particularly appropriate at this stage to explore with students some of the ideas for effective and respectful group discussion found in the Tool Enhancement Supplement, specifically the *Tool Enhancement: Consensus decision making* (Blackline Master E).

Tool Enhancement: negotiating agreement

Preparing final demands (45 - 90 minutes)

◆ In light of the information gained from investigation of the other colonies, direct each delegation to reconsider the list of conditions in its Preliminary Proposal. Remind students of two objectives in re-drafting their conditions: they are to specify the conditions or terms that would have to be offered to their colony in order for it to join the federal union and, in order to be considered by Great Britain for implementation in the final bill, that would be acceptable to the other colonies as well.

◆ To assist students in thinking about their final conditions, introduce ideas found in the Tool Enhancement Supplement, specifically Activity D: Role Reversal and Activity E: Refining Arguments and Counter-arguments.

Tool Enhancement: developing arguments

◆ Distribute to each delegation two copies (for a working and a polished draft) of *Student Activity: Final proposal* (Blackline Master #19) upon which to record its final set of conditions and accompanying justifications. Explain to students that the proposal should indicate the following:

- those conditions that have not changed from the Preliminary Proposal and the justifications for retaining these demands;
- any modifications, additions or deletions in demands and the justification for each change.

Remind students that their final set of conditions must be historically plausible, thoughtfully address their colony's needs, and be clearly and specifically stated. As well, the proposal must show sensitivity to the needs of the other colonies. Each delegation is to submit a polished copy of their Final Proposal to the teacher.

◆ As in the Preliminary Proposal stage, assign each delegation to prepare and rehearse a five-minute presentation which outlines its final demands and their justification. Remind students that all members of each delegation must participate meaningfully in the presentation, that the presentation should make effective use of visual aids, that it should be clearly structured and that delivery should be audible and clear. Encourage each delegation to critique its oral presentation prior to actual delivery to the Conference as suggested in Activity I: Productive Peer Critique found in the Tool Enhancement Supplement.

Tool Enhancement: presenting ideas

Presenting final proposals (45 - 60 minutes)

- Arrange for each delegation to make its five-minute Final Proposal presentation. Follow this with an opportunity for each colony to make two-minute closing commentaries should they wish. These comments may be reactions to the presentations given, final attempts for compromise in the form of new concessions, reiterations of key demands or simply remarks on the negotiation process.

- OPTIONAL: For presentation of final proposals and voting, assign students to role play a representative of his/her colony in costume. Students could also research and write mini-biographies of their assigned representative for a more accurate role play.

- Use the rubrics found in *Assessment: Final proposals* (Blackline Master #28) to assign a group mark to the written and oral components of each colony's final proposals.

final proposals

Voting on the Bill

- The voting stage has two parts: (1) examination of the Confederation bill and (2) justification of the decision. Referring to the earlier mentioned *Support Material: Conference agenda* (Blackline Master #22), outline for the class these two tasks.

Examining the proposed bill (30 minutes)

- The teacher, acting as the Legislature of Great Britain/Governor General, creates a bill based on the Final Proposals and any pertinent closing comments presented in Round Two. The goal is to create a bill that will please as many colonies as possible, using only the recommendations proposed by the colonies, unless an obvious compromise of the various positions can be made. Drafting of this bill may take up to one hour. The scroll on the following page is an example of a Confederation bill drafted for a secondary class in Vancouver.

- As the Queen's representative in Canada, present to each delegation several copies of the Confederation bill, drafted using *Support Material: Confederation Bill* (Blackline Master #25) as the template. Since no further negotiation is possible, requests to change the bill cannot be considered. Each colony is given one vote.

- Allow the delegations time to deliberate on whether or not to vote for the bill; in other words, to accept or reject Canadian Confederation. To facilitate their deliberations, distribute to each delegation two copies (for a working and a polished draft) of *Student Activity: Decision on the Confederation Bill* (Blackline Master #20). To begin, assign each delegation to complete the first two sections of this sheet by indicating

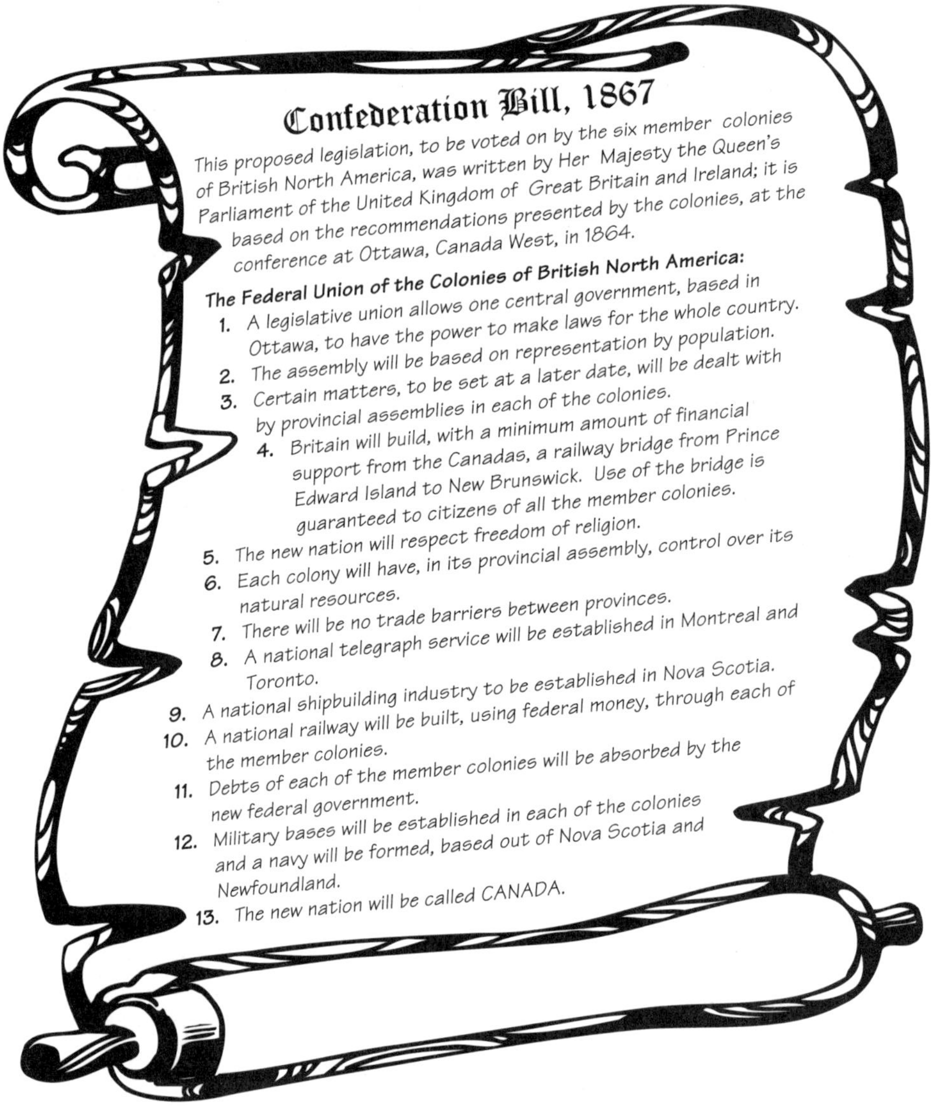

Confederation Bill, 1867

This proposed legislation, to be voted on by the six member colonies of British North America, was written by Her Majesty the Queen's Parliament of the United Kingdom of Great Britain and Ireland; it is based on the recommendations presented by the colonies, at the conference at Ottawa, Canada West, in 1864.

The Federal Union of the Colonies of British North America:

1. A legislative union allows one central government, based in Ottawa, to have the power to make laws for the whole country.
2. The assembly will be based on representation by population.
3. Certain matters, to be set at a later date, will be dealt with by provincial assemblies in each of the colonies.
4. Britain will build, with a minimum amount of financial support from the Canadas, a railway bridge from Prince Edward Island to New Brunswick. Use of the bridge is guaranteed to citizens of all the member colonies.
5. The new nation will respect freedom of religion.
6. Each colony will have, in its provincial assembly, control over its natural resources.
7. There will be no trade barriers between provinces.
8. A national telegraph service will be established in Montreal and Toronto.
9. A national shipbuilding industry to be established in Nova Scotia.
10. A national railway will be built, using federal money, through each of the member colonies.
11. Debts of each of the member colonies will be absorbed by the new federal government.
12. Military bases will be established in each of the colonies and a navy will be formed, based out of Nova Scotia and Newfoundland.
13. The new nation will be called CANADA.

- those issues that have been resolved by the proposed bill, and the likely benefits to its colony following from this;
- those issues that have not been resolved by the proposed bill, and the likely negative consequences following from this.

Justifying the decision (30 minutes)

◆ You may want to teach, or to remind students about, a few of the tools related to reaching consensus before setting them to the next task. At the very least, explain what it means to reach a decision by consensus. The strategies outlined in the Blackline Master *Consensus decision making* found in the Tool Enhancement Supplement would be especially useful.

Tool Enhancement: negotiating agreement

◆ Each delegation should try to reach consensus based upon its decision to accept or reject the bill. The decision and the key reasons for this decision should be recorded on the bottom part of *Student Activity: Decision on the Confederation Bill* (Blackline

Master #20). The polished copy of this document should be submitted for evaluation to the teacher. Encourage delegations to keep their final decision to themselves until the formal vote is take.

- Organize a final voting session. Explain to students that only those colonies voting for Confederation join the new union. Explain also that if there are colonies which do not vote in favour of the bill, it may be necessary to have another Confederation Conference attended by signing members of the bill in order to work out any changes or problems caused by the absence of the non-signing colonies. Ask for one representative from each colony to register his or her colony's vote at the front of the classroom. Should the bill be passed, organize a signing session of the *Confederation Bill* (Blackline Master #25). Ratify the bill immediately with the signature of the teacher (representing Great Britain) and give Royal Assent to the British North America Act.

- Use the rubric found in *Assessment: Final decisions* (Blackline Master #29) to assign a group mark to the *Decision on the Confederation Bill* (Blackline Master #20).

final decisions

Debriefing

- The debriefing stage has two parts: (1) reflections on the experience and (2) peer- and self-assessment. Referring to the earlier mentioned *Support Material: Conference agenda* (Blackline Master #22) outline for the class these two tasks.

Reflecting on the experience (30 - 45 minutes)

- Once the simulation is completed, invite students to reflect on the lessons to be learned from this exercise. Assign students to complete *Student Activity: Reflecting on the experience* (Blackline Master #21) and then organize a class discussion based on their responses. (Alternatively, students can be asked to complete these surveys anonymously and submit them to the teacher.)

- OPTIONAL: In lieu of using *Student Activity: Reflecting on the experience* (Blackline Master #21), after discussing students' reactions to the simulation in class, assign students to consolidate in three paragraphs their reflections on the following topics:
  - their personal experiences and feelings as a participant in this simulated exercise;
  - observations about the difficulties that were encountered in reaching the actual agreement on Confederation;
  - suggested strategies for negotiating agreements on personal and on national matters.

Peer- and self-assessment (30 - 45 minutes)

- To assess students' individual contributions to their group, use the peer- and self-assessment procedure described in *Assessment: Individual contributions* (Blackline Master #30). You may prefer to generate two marks from this procedure—students' self-assessed mark and the average of their peers' assessments of them—or simply to average all assessments, including their self-assessment, into one mark. If students are new to peer- and self-assessment, stress the importance of students providing thoughtful and fair assessments of themselves and of each other. One way to increase the likelihood of students being conscientious and fair is to assess their assessments, see *Assessment: Assessing student assessments* (Blackline Master #31).

individual contributions

assessing student assessments

Extension

- Provide students with a simplified summary of the actual British North America Act in order to compare their bill with the one approved in 1867.
- Assign students to write an essay speculating on what Canada would be like today if the British North America Act they signed actually became law.
- Assign students to write a newspaper article describing the events of the Confederation Conference.
- Organize a Confederation tea party, asking each colony to bring something that would have been served at a tea in 1867. The menu might include tea sandwiches, tea biscuits and scones.

Blackline Masters

Briefing Sheets

1 The Ottawa Conference
2 Canada East—Background
3 Canada East—Viewpoints
4 Canada West—Background
5 Canada West—Viewpoints
6 New Brunswick—Background
7 New Brunswick—Viewpoints
8 Newfoundland—Background
9 Newfoundland—Viewpoints
10 Nova Scotia—Background
11 Nova Scotia—Viewpoints
12 Prince Edward Island—Background
13 Prince Edward Island—Viewpoints
14 Maps of the colonies in 1864
15 Annotated bibliography on Confederation

Student Activities

16 Father of Confederation profile
17 Colonial perspective
18 Preliminary proposal
19 Final proposal
20 Decision on Confederation bill
21 Reflecting on the experience

Support Materials

22 Conference agenda
23 Conference materials
24 Representatives' name tags
25 Confederation Bill, 1867

Assessment

26 Knowledge of colonial perspectives
27 Preliminary proposals
28 Final proposals
29 Final decisions
30 Individual contributions
31 Assessing student assessments

The Ottawa Conference

Introduction

In the 1860s the possibility of Confederation of the British North American colonies—Canada East and Canada West, New Brunswick, Nova Scotia, Newfoundland and Prince Edward Island—meant different things to different people. For some, Confederation was British North America's only chance for political and economic survival, and for others, such a union would give each of the regions an opportunity to maintain an individual identity, while, at the same time, reap the economic and political benefits a large nation brings. For Great Britain, a union of the colonies would lift a much needed burden off its shoulders. Even so, there were many people in each of the colonies who vehemently opposed any federal union. Ultimately, compromise and sophisticated deal making over the course of several years and three separate conferences, allowed some of the colonies to come together to form a new country called Canada.

Critical Challenge

We will attempt to form a country by participating in the Ottawa Conference, a fictitious Confederation conference set in 1864. The class will be divided into six groups, each representing a colony at the conference. The conference consists of two rounds of negotiations and presentations during which time the colonies will attempt to draft an agreement for Canadian Confederation. At the end of these negotiations, the teacher, who will be acting as the Governor General, will draft a bill for Confederation. Each colony will then determine if this bill meets its demands and decide whether or not to sign for Confederation.

- **Preliminary Proposals**. Each delegation studies the Briefing Sheets on its colony and conducts library research in order to complete a Colonial Perspective profile. This document should serve as the backbone for the colony's Preliminary Proposal, which is a set of that colony's requirements for joining Confederation. Every colony will provide written copies of their Preliminary Proposal to the other colonies. Each delegation will create and practice a five-minute class presentation identifying and defending its requirements. All members of the delegation must be involved in a meaningful way in the presentation. The presentation must make effective use of visual aids. At the end of each presentation, each of the other delegations must ask the presenting colony one question about a pertinent issue.
- **Final Proposals.** Each delegation will regroup to discuss the Preliminary Proposals from the other colonies. Each colony member now becomes an expert on one other colony. After reading the Briefing Sheet and other documents for that colony, each "expert" completes a Colonial Perspective and then briefs his or her delegation on the colony. Each delegation should revise its Preliminary Proposal by developing a Final Proposal that addresses the vital concerns of its colony and of the other colonies. Each delegation will prepare and practice a five-minute class presentation identifying and defending its final requirements. All members of the delegation must be involved in a meaningful way in the presentation. The presentation must make effective use of visual aids. After the presentations, each colony may make a two-minute closing statement to the Conference.
- **Voting on the Bill.** The teacher, acting as the Governor General, presents the colonies with a Confederation bill based on the final proposals presented at the Ottawa Conference. This Confederation bill will outline the terms and conditions of the new federal union. (Each colony will have one vote on whether or not to sign the bill; in other words, to accept Canadian Confederation.) Each delegation must deliberate and try to reach a consensus decision. Only those colonies voting for the bill will form Canada.

Evaluation

A large part of the evaluation for the simulation will be group marks for the quality of the oral and written presentations. Students will be assessed individually by their fellow students on how much they have contributed to these group projects and on how well they have worked with one another.

Presentation of Preliminary Proposal (group mark)	40	marks
Presentation of Final Proposal (group mark)	40	marks
Decision on Confederation (group mark)	25	marks
Knowledge of Colonial Perspectives (individual mark)	25	marks
Peer- and self-assessment (individual mark)	20	marks
Teacher assessment of student assessments (individual mark)	10	marks
TOTAL	**160**	**marks**

Canada East – Background

Population (circa 1860): 1,112,000

Urban centres: Montreal (107,225)
Quebec (59,700)

Key figures: George-Etienne Cartier (1814-1873)
Alexander T. Galt (1817-1893)
Thomas D'Arcy McGee (1825-1868)
Hector L. Langevin (1826-1906)
Etienne P. Taché (1795-1865)
Jean Charles Champais (1811-1885)

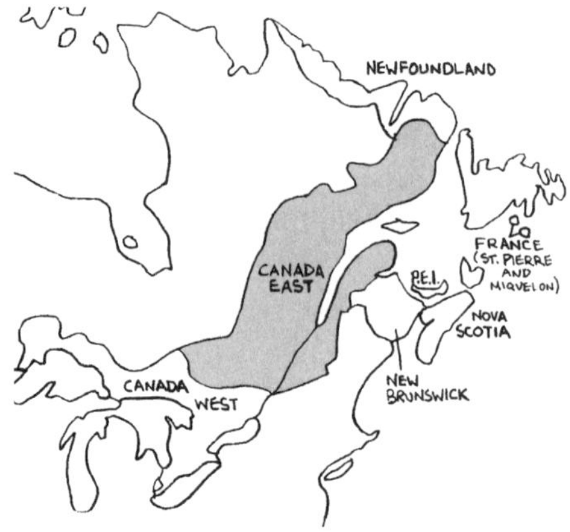

By the 1860s, the Province of Canada (encompassing both Canada East and Canada West) is the most populous, the largest in size and the most powerful of the British North American colonies. Canada East is dominated by the lumbering industry and an agricultural economy. The colony's urban centre, Montreal, is the most populous city in British North America. It is, in fact, almost double the size of Quebec, the second largest city in the colonies. The wealthy lumber merchants are central members of the city's elite, along with a rising group of industrialists: owners of iron and steel plants, flour mills and steamship lines. This group, dominated by Scotsmen, have built grand mansions along the slopes of Mount Royal, in the centre of the city. Though English speaking Protestants make up only 15 percent of the colony's population, they dominate the commercial and political life of the colony.

It is the rural habitant, however, the French Canadian farmer, who makes up the bulk of the rural population. Living, by and large, along the shores of the St. Lawrence River, the French speaking, Roman Catholic habitant lives a traditional way of life, producing potatoes, rye, buckwheat and livestock.

A serious problem, magnified by the recent growth in Canada West, is the political deadlock in the Canadian Parliament. The inability to form a majority government led to three different administrations between 1861 and 1864. In 1864, however, the Great Coalition was formed. Made up of Conservatives, Clear Grits and Reformers from Canada East and Canada West, the Great Coalition called for, among other things, a federal union of the British North American colonies. Many feared the destruction of French culture in any union of the British North American colonies. But George-Etienne Cartier, the French Canadian member of the Coalition, believed that only in a federal union of the colonies would French Canadian culture survive and, in turn, flourish. In the new federal union, he argued, French Canadians would still control all matters concerning language, religion, civil law and education within the province.

The Grand Trunk is the colony's central railway line. With its headquarters in Montreal, and an impressive network of lines (including the Victoria Bridge, which spans the St. Lawrence and is the world's longest bridge), transportation is still hampered by two factors. Not only does the colony lack a year-round, ice-free port, but also one cannot travel from Canada East to the Maritime colonies without travelling through the United States. For six months of the year, Canadian imports and exports are carried on American railways, on American soil and, often, shipped in and out of American ports. The solution, in many Canadian minds, is an Intercolonial Railway. If the Intercolonial were built, it would run through New Brunswick, Nova Scotia and end in Halifax. Canada East's commercial potential would be immense.

The Intercolonial Railway would also help in the defence of British North America. Great Britain's reluctance to defend the colonies has made closer ties between the British North American colonies crucial to security. The threat of Fenian raids along the Canada-United States border near Montreal, as well as the threat of American invasion during the Civil War, make an Intercolonial Railway necessary for mobilizing troops.

Canada East – Viewpoints

Drawn from R.C. Brown & M.E. Prang (Eds.), *Confederation to 1949* (Prentice, 1966); J.M. Bliss (Ed.), *Canadian History in Documents* (Ryerson, 1966); H.H. Herstein, L.S. Hughes & R.C. Kirbyson, *Challenge and Survival: The History of Canada* (Prentice, 1970); J.S. Reid, K. McNaught & H.S. Crowe, *A Source-Book of Canadian History* (Longmans, 1959) and P.B. Waite, *Confederation, 1854-1867* (Holt, 1972).

1. **Thomas D'Arcy McGee's vision of a new Northern Nation** (1860)

   I have spoken with a sole single desire for the increase, prosperity, freedom and honour of this incipient Northern Nation. I call it a Northern Nation—for such it must become, if all of us do our duty to the last . . . I see in the not remote distance one great nationality, bound, like the shield of Achilles, by the blue rim of Ocean. I see it quartered into many communities, each disposing of its internal affairs, but all bound together by free institutions, free intercourse and free commerce. I see within the round of that shield the peaks of the Western Mountains and the crests of the Eastern waves, the winding Assiniboine, the five-fold lakes, the St. Lawrence, the Ottawa, the Saguenay, the St. John, and the basin of the Minas. By all these flowing waters in all the valleys they fertilize, in all the cities they visit in their courses, I see a generation of industrious, contented, moral men, free in name and in fact—men capable of maintaining, in peace and in war, a constitution worthy of such a country!

2. **French Canadian distrust of Confederation**
   (*L'Ordre* Montreal, 4 May 1860); *translation*

   What French Canadian has not in his heart cursed a hundred times the Union of the two Canadas? . . . Others have wanted in turn to *anglicize* us and *protestantize* us: after a century of ignoble hopes and base efforts, convinced of their failure, they now want to destroy our constitution. . . . What would Upper Canada be today without the Union? Nothing more or less than a forest put up for auction by British capitalists to repay their investments. The only solution is repeal of the Union. Upper Canada does not like living with us: we like it less.

3. **Provincial rights in the Union**
   (*Le Courrier du Canada* Quebec, 10 October 1864); *translation*

   Let us give to each province its own distinct autonomy, let each province be master in its own house in matters of social organization, ownership of public property, preservation of its language, laws and institutions, while protecting minorities everywhere, and let us unite all parts into a federal agreement covering matters in which a common defense and common interests see us all joined on the same ground.

4. **George-Etienne Cartier**
   (Canadian Parliament, Winter 1865)

   (I)f union were attained, we would form a political nationality with which neither the national origin, nor the religion of any individual, would interfere . . . with regard to the objection based on this fact, to the effect that a great nation could not be formed because Lower Canada was in great part French and Catholic, and Upper Canada British and Protestant, and the Lower Provinces were mixed, it was futile and worthless in the extreme. Look, for instance, at the United Kingdom, inhabited as it was by three great races. Had the diversity of race impeded the glory, the progress, the wealth of England? Had they not rather contributed their share to the greatness of the Empire? . . . In our own Federation we should have Catholic and Protestant, English, French, Irish and

Scotch, and each by his efforts and his success would increase the prosperity and glory of the new Confederacy.

5. **Joseph Perrault**
(Canadian Parliament, Winter 1865)

(W)ith Confederation, as we shall be in the great minority in the General Parliament, which as all the important powers in relation to legislation, we shall have to carry on a constant contest for the defence and preservation of our political rights and of our liberty.

6. **D'Arcy McGee on the American threat**
(Canadian Parliament, 9 February 1865)

These are frightful figures [U.S. military] for the capacity of destruction they represent, for the heaps of carnage that they represent, for the quantity of human blood spilt that they represent, for the lust of conquest that they represent, for the evil passion that they represent, and for the arrest of the onward progress of civilization that they represent. . . . They [the Americans] coveted Florida, and they seized it; they coveted Louisiana, and purchased it; they coveted Texas and stole it; and then they picked a quarrel with Mexico, which ended by their getting California. . . . had we not the strong arm of England over us, we would not now have had a separate existence.

7. **Alexander T. Galt on the need for Intercolonial trade**
(Canadian Parliament, 7 February 1865)

Intercolonial trade has been, indeed, of the most insignificant character; we have looked far more to our commercial relations with the neighbouring—though a foreign country—than to the interchange of our own products, which would have retained the benefits of our trade within ourselves; hostile tariffs have interfered with the free interchange of the products of the labour of all the colonies, and one of the greatest and most immediate benefits to be derived from their union, will spring from the breaking down of these barriers and the opening up of the markets of all the provinces to the different industries of each.

8. **Hector L. Langevin on the fate of French Canadians in Confederation**
(Canadian Parliament, Winter 1865)

But what would be the fate of the French Canadians in the case of annexation to the United States? Let us profit by the example of the French race in the United States, and enquire what has been the fate of the French in Louisiana? What has become of them? What has become of their language, their customs, their manners and their institutions? After the war, hardly a trace will remain to show that the French race has passed that way . . . we live in peace at the present day and are perfectly comfortable; Catholics and Protestants have the same rights and religious liberty, and they live peacefully together as there was but one religion in the land.

Canada West – Background

Population (circa 1860): 1,396,000

Urban centres: Toronto (56,000)
Hamilton (26,700)
Ottawa (21,500)
London (15,826)
Kingston (12,400)
Brantford (8,100)

Key figures: John A. Macdonald (1815-1891)
George Brown (1818-1880)
William McDougall (1822-1905)
Alexander Campbell (1822-1892)
Oliver Mowat (1820-1903)
James Cockburn (1819-1883)
William P. Howland (1811-1907)

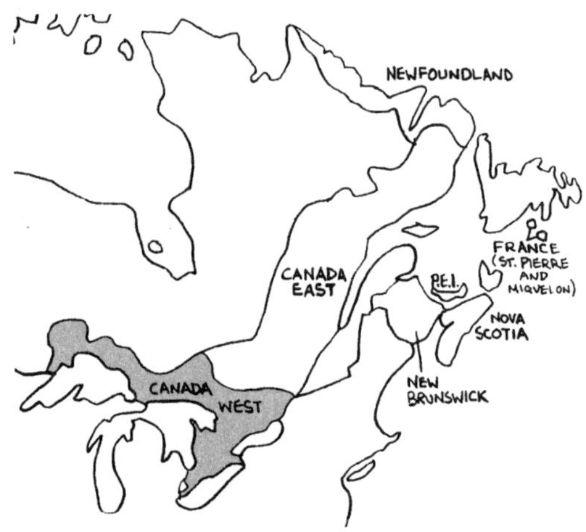

By the 1860s, the Province of Canada (encompassing both Canada East and Canada West) is the most populous, the largest in size, and the most powerful of the British North American colonies. Overwhelmingly Protestant and English speaking, Canada West's population is now greater than Canada East's and it is growing at a faster rate. As a result, it is becoming increasingly difficult to find good farm land in a colony where over 80 percent of its residents live in the countryside. To prevent young farmers from moving to the American West, and to halt American expansion into the British North West Territory, many politicians and newspapermen—led by George Brown and his Toronto *Globe*—are calling for Canada to acquire, to settle and, ultimately, to govern the Western lands, presently controlled by the Hudson's Bay Company.

Another problem magnified by the recent growth is the political deadlock in the Canadian Parliament. The inability to form a majority government led to three different administrations between 1861 and 1864. In 1864, however, the Great Coalition was formed. Made up of Conservatives, Clear Grits and Reformers from Canada West and Canada East, the Great Coalition called for a federal union of the British North American colonies, "rep by pop" (representation by population), and the incorporation of the North West Territory into any Confederation. "Rep by pop," the central demand of George Brown's Clear Grits, would give Canada West the political power its increasing population deserves.

Agriculture and lumbering are the chief industries in Canada West. The wealthy wheat and lumber merchants form the core of the colony's elite. The key to Canada West's economy is the Grand Trunk Railway. The colony's already extensive rail network has made Toronto the commercial centre. Goods from all regions of the colony flow into the growing city, ready for export or to be shipped elsewhere within Canada West.

Trade, however, is hampered by two factors. Not only does the colony lack a year-round, ice-free port, but also one cannot travel from Canada West to the Maritime colonies without travelling through the United States. For six months of the year, Canadian imports and exports are carried on American railways, on American soil and, often, shipped in and out of American ports. The solution, in many Canadian minds, is an Intercolonial Railway. If the Intercolonial were built, it would run through Canada East, New Brunswick, Nova Scotia and end in Halifax. Canada West's commercial potential would be immense.

The Intercolonial Railway would also help in the defence of British North America. Great Britain's reluctance to defend the colonies has made closer ties between the British North American provinces crucial to security. The threat of Fenian raids along the Canada-United States border in the Niagara region and along the St. Lawrence River, as well as the threat of American invasion during the Civil War, make an Intercolonial Railway necessary for mobilizing troops.

Canada West – Viewpoints

Drawn from R.D. Francis, R. Jones & D.B. Smith, *Origins: Canadian History to Confederation* (Holt, 1988); H.H. Herstein, L.J. Hughes & R.C. Kirbyson, *Challenge and Survival: The History of Canada* (Prentice, 1970) and P.B. Waite, *Confederation, 1854-1867* (Holt, 1972).

1. **Opposition to the Intercolonial Railroad**
   (Toronto *Globe*, 8 October 1858)

   The commercial advantages to be derived from a union of the Lower Provinces [the Maritimes] are hardly appreciable, while in the boundless west there lies open to us a field of enterprise which might cause wealth to flow into every city and village of our land. Why should Canada, at this moment, spend any portion of her means in building a road to Halifax? She will not . . . find in the Lower Provinces a market for any of her manufactures. As an outlet to the ocean, this intercolonial railroad is a mere farce. No one able to take ship at Portland [Maine, U.S.A.] or Quebec [City], would ever dream of travelling by railway to Halifax. And yet this work will engage four or five millions [£] of capital. If the Imperial Government [Great Britain] is willing to grant assistance for the development of British power in North America, let her grant it in aid of the Pacific Railway. . . . Let her [Britain] expend it in founding a great colony on Lake Winnipeg and the Saskatchewan.

2. **The importance of the Intercolonial**
   (Toronto *Leader*, 25 October 1862)

   A great country such as Nature has destined this to be would not be justified in refusing to acquire a winter sea-port, when the object can be obtained upon reasonable terms. Without it, what is the possible future of Canada? A back country, with no access to the seaboard, during six months of the year, but through the territory of a foreign power, occupies a position of deplorable dependence.

3. **Opposition to the Intercolonial**
   (St. Thomas *Weekly Dispatch*, 13 November 1862)

   What possible benefit can Canada derive from developing the resources of Nova Scotia, New Brunswick or the North West Territory. If the Intercolonial Railway was going to cut through rich portions of the fertile lands of Canada, like the Grand Trunk . . . there might be some reason to talk [as McGee does] of further development; but when it can only push our produce some five or six hundred miles further by rail through a barren country, when we can now send it by sea or by Portland [Maine, U.S.A.], and thereby make a commercial mart of Halifax instead of Montreal or Quebec, such development becomes mere delusion.

4. **Support for federation**
   (Oshawa *Vindicator*, 31 August 1864)

   No system of Government could be fairer, or could be better calculated to give satisfaction to all parties interested, than the federal system.

5. **On American Aggression**
   (Hamilton *Evening Times*, 18 March 1865)

   Are we going too far when we say there does not exist in the world a country more ineligible for defensive purposes than Canada? We are all frontier, and are open to attack at almost every point of that frontier. . . . Inherently strong, and armed and equipped at every point, the United States stands forth a giant encased in armor. What would our prospects be should we come into collision with this power?

6. **John A. Macdonald for the Intercolonial Railway**
(A speech in Halifax, 12 September 1864)

(T)his railway must be a national work, and Canada will cheerfully contribute to the utmost extent in order to make that important link, without which no political connection can be complete. What will be the consequence to this city, prosperous as it is, from that communication? Montreal is at this moment competing with New York for the trade of the great West. Build the road and Halifax will become one of the great emporiums of the world. All the great resources of the west will come over the immense railways of Canada to the bosom of your harbour.

7. **George Brown on the economic advantages of Confederation**
(Canadian Parliament, 8 February 1865)

I go heartily for the union, because it will throw down the barriers of trade and give us the control of a market of four millions of people. What one thing has contributed so much to the wondrous material progress of the United States as the free passage of their products from one State to another? . . . I am in favour of a union of the provinces . . . because it will make us the third maritime state of the world. When this union is accomplished, but two countries in the world will be superior in maritime influence to British [North] America—and those are Great Britain and the United States.

8. **Marching Song of the Fenian Brotherhood**
(From sometime in the 1860s)

The group, formed in the United States in 1859, was dedicated to fighting for the independence of Ireland.

We are the Fenian Brotherhood,
skilled in the art of war,
And we're going to fight for Ireland,
the land that we adore,
Many battles have we won, along with
the boys in blue,
And we'll go and capture Canada for
we've got nothing else to do.

New Brunswick – Background

Population (circa 1860): 252,000

Urban centre: Saint John (28,805)

Key figures:
Samuel Leonard Tilley (1818-1896)
John M. Johnson (1818-1868)
William H. Steeves (1814-1873)
Edward B. Chandler (1800-1880)
John Hamilton Gray (1814-1889)
Peter Mitchell (1824-1899)
Charles Fisher (1808-1880)
R.D. Wilmot (1809-1891)

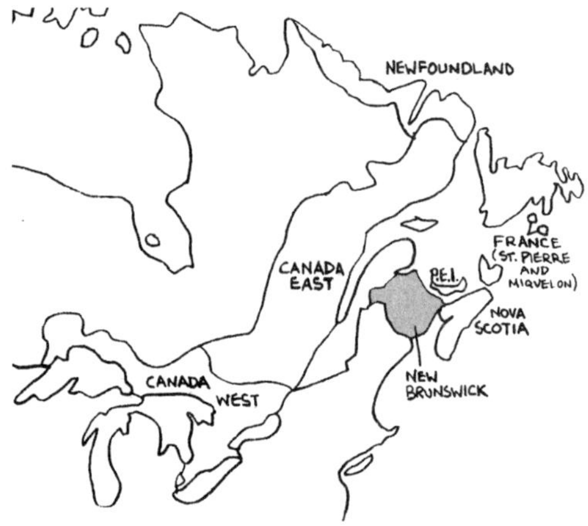

Since the early nineteenth century, New Brunswick life has been dominated by the timber trade. The economy and even the colonial character are shaped by it. The vast forests in the western part of the colony are the bedrock of this industry and, in many ways, form a natural barrier between New Brunswick and the rest of the continent. In fact, only three to four percent of the colony's trade is with the Canadas, while twenty percent of all British timber imports come from New Brunswick. This rich supply of lumber has led to a thriving shipbuilding industry in Saint John. And while people in the colony are also involved in farming and fishing, any significant growth of an agricultural economy is impeded by the attractive profits in the lumber industry.

Though it is not the colony's capital, Saint John is one of the Maritimes' largest cities. Deeply influenced by the rugged life of the timber trade, Saint John is known as a "fast city," with lively and aggressive residents. Fredericton, a sleepy town by comparison, has been chosen the colonial capital because it is upriver and easier to defend in case of attack.

The fear of attack from the United States army, or the Fenian marauders, is very real. The colony's long border with the United States is poorly defended and not easily accessible to troops from outside New Brunswick. British troops sent to defend the colonies during the winter months of the American Civil War had to travel by sled through New Brunswick. Britain's reluctance to defend the colonies in the future has made closer ties between the British North American provinces crucial to security.

The solution to any questions surrounding colonial defence is an Intercolonial Railway. Though it would be too costly for New Brunswick to pay for on its own, the rail line would be central to the colony's defence and, if built, would open up New Brunswick economically to both the Western and the Eastern North American markets. Saint John would now be connected to Montreal in the West and Halifax in the East.

Briefing Sheet

6

New Brunswick – Viewpoints

Drawn from R.D. Francis, R. Jones & D.B. Smith, *Origins: Canadian History to Confederation* (Holt, 1988); H.H. Herstein, L.J. Hughes & R.C. Kirbyson, *Challenge and Survival: The History of Canada* (Prentice, 1970) and P.B. Waite, *Confederation, 1854-1867* (Holt, 1972).

1. **Marching song of the Fenian Brotherhood**
(From sometime in the 1860s)

The group, formed in the United States in 1859, was dedicated to fighting for the independence of Ireland.

> We are the Fenian Brotherhood,
> skilled in the art of war,
> And we're going to fight for Ireland,
> the land that we adore,
> Many battles have we won, along with
> the boys in blue,
> And we'll go and capture Canada for
> we've got nothing else to do.

2. **On nationhood**
(St. John *Morning News*, 2 June 1858)

The experience of history tells us that Colonies cannot be Colonies forever; and Canada in time will assume that character among the Nations of the world which her position, her wealth, and her intelligence will entitle her to.

3. **On Union and the Intercolonial Railway**
(Saint John *Morning News*, 25 August 1858)

We regard the building of the inter-Colonial Railroad to be an absolute condition [of British North American union]—for there can be no union of feeling and sentiment, or harmony of understanding and action, while such physical difficulties are geographically described, stand in the way of the people of Canada West and those of Nova Scotia, separated by a distance of 800 miles, and who can only communicate with one another through American territory.

4. **Britain's view of British North America**
(St. John *Morning Telegraph*, 12 September 1864)

(T)he Mother Country [Great Britain] is becoming tired of her Colonial dependencies. This is undoubtedly the prevailing sentiment among the English Statesmen. . . . It does not pay at present to retain the British North American Colonies. . . . Why, then, should the British take any special interest in our welfare? Blood goes for nothing these days, for Anglo-Saxon blood is everywhere. Commerce is everything—and everything must succumb to commercial calculations.

We consider, then, that the time is near at hand when we shall be told to shift for ourselves; and the American difficulty, if we are not greatly mistaken, has brought the time considerably nearer to our doors than it otherwise would be.

Briefing Sheet

7

Newfoundland – Background

Population (circa 1860): 122,000

Urban centre: Saint John's (30,475)

Key figures: Frederick B.T. Carter (1819-1900)
Ambrose Shea (1815-1905)

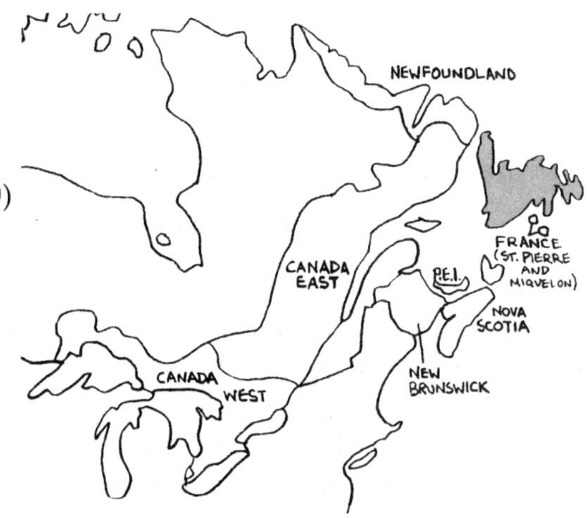

Newfoundland is largely unknown to the other British North American colonies, especially the Canadas. Without Labrador, it is nearly the size of the other three Maritime colonies put together. Most of the island is uninhabited. Its coasts are charted, but the island's interior—consisting of dense forest, swamps, rock, without roads of any kind—remains unmapped. The land is so rugged that fewer than one hundred square miles are cultivated.

The colony's population settled in the southeast corner of the island. St. John's, the colony's capital and urban centre, is at the easternmost point of North America. These geographical factors make Newfoundland's link with Britain—to the east across the Atlantic—more immediate than its ties to North America. From its earliest days, Newfoundland has traded fish with Britain, in exchange for manufactured goods. The colonists wear British-made clothes, read British newspapers and magazines and, for many, London feels closer than Ottawa.

Fishing and seal hunting, traditionally the backbone of the Newfoundland economy, have been virtually nonexistent for years. Without these staple industries, many Newfoundlanders are nearly destitute. Debt is common among fishermen, and one-third of all money generated in Newfoundland is used to help the poor.

Britain's new attitude towards the colonies has left many Newfoundlanders worried. Since the early eighteenth century, French fishermen have caught fish in Newfoundland's waters and dried them on the island's west shore. Newfoundlanders suspect that the French would like to use this land for more than drying fish. Newfoundland relies on the British navy for protection. Though Newfoundland does not directly border on the United States and there is no threat of an American attack, Britain's reluctance to provide colonial defence in the future worries Newfoundland. As an island, a rail link with the other British North American colonies is irrelevant. But without the support of the British navy Newfoundlanders wonder how they will defend themselves.

Briefing Sheet

8

Newfoundland – Viewpoints

Drawn from P.B. Waite, *Confederation, 1854-1867* (Holt, 1972) and P.B. Waite, *The Life and Times of Confederation, 1864-1867: Politics, Newspapers and the Union of British North America* (University of Toronto Press, 1962).

1. **Anti-Union**
(St. John's *Patriot*, 29 November 1864)

   We have no faith in a political union with Canada at all. We deem it—"Better to endure the ills we have, than flee to others that we know not of."

2. **Complexity of Confederation**
(St. John's *Patriot*, 6 December 1864)

   Nor can we duly comprehend the magnitude of being the contemptible fag-end of such a compact. . . . This Federal scheme is an after-thought of the Canadians. We cannot deny that it is a brilliant thing on paper—this Confederation this Great United British America, which shall reach from "Newfoundland to the Rocky Mountains" but its brilliancy does not dazzle us as to the duties which must necessarily fall to our share. . . . That TAXATION for all local purposes will be resorted to, is as plain as ABC.

3. **The need for change**
(St. John's *Newfoundlander*, 5 January 1865)

   If any of the Provinces more than another should seek a change, it is this. We do not mean to assert that we should adopt a change blindly, but unlike our Sister Colonies, our circumstances—the condition of our Trade—the depressed state of our people, demand a change, even if Confederation had never been proposed.

4. **Caution**
(Harbour Grace *Standard*, 7 December 1864)

   Confederation of the British North American Colonies seems to be the all-absorbing question just at present. . . . This is certainly a matter of vital importance to this country, and should be approached with the greatest caution by the people of Newfoundland.

Briefing Sheet

9

Nova Scotia – Background

Population (circa 1860): 331,000

Urban Centre: Halifax (29,580)

Key figures: Charles Tupper (1821-1915)
William A. Henry (1816-1888)
Jonathon McCully (1809-1877)
Adams G. Archibald (1814-1892)
Robert B. Dickey (1811-1903)
John William Ritchie (1808-1890)

Nova Scotia plays a key role in sea communications between Great Britain and the United States. It is common for ships bound for New York and Boston to stop at Halifax and for ships bound for Liverpool and London to also dock here. For many seafaring Nova Scotians, the British West Indies feels closer than Quebec. Above all, Halifax is the central base for the British Navy in the West Atlantic. With an ice-free harbour and the citadel, a grand fortress on a hill overlooking the harbour and the city, Halifax stands out as a symbol of Imperial Britain's power.

A great majority of Nova Scotians make their living, in one way or another, from the ocean. Almost half of the colony's residents are fishermen, while others work on ships exporting cod and importing coal and wool. Over one-third of Nova Scotia's exports are fish and fish products. Shipbuilding is an enormously prosperous industry, and Nova Scotians produce and own more ships than residents in any other British North American colony.

Politics is highly sophisticated in Nova Scotia. The first British North American colony to achieve responsible government, in 1848, Nova Scotian politics is marked by its leaders. First, Joseph Howe, and later, Charles Tupper, brought unity to the colony by bringing together various regional and cultural interests. A federal union, though not initiated by the Nova Scotians, is something to be considered, discussed and debated.

The significant British military and naval presence in Halifax reduces the immediate threat of an American attack. But Great Britain's reluctance to defend the colonies in the future makes closer ties between the British North American provinces crucial to security. Any change in British policy would have a serious effect on Nova Scotia because of this close relationship.

There is no railway connection between Nova Scotia and the Canadas. Many prominent British North Americans believe an Intercolonial Railway must end in Halifax. The Intercolonial would not only help in the defence of British North America, but it would also open up other North American markets to Nova Scotian businesses. Since the 1850s Nova Scotia has relied on reciprocity (free trade) with the United States. It is crucial to Nova Scotia's future to secure other North American markets, in case the United States imposes tariffs on Nova Scotian goods.

Briefing Sheet **10**

Nova Scotia – Viewpoints

Drawn from D. Creighton, *Dominion of the North: The History of Canada* (Macmillan, 1957); H.H. Herstein, L.J. Hughes & R.C. Kirbyson, *Challenge and Survival: The History of Canada* (Prentice, 1970) and P.B. Waite, *Confederation, 1854-1867* (Holt, 1972).

1. **Anti-Confederate on Nova Scotia's natural connection to the United States** (1865)

   Look at the geographical position of this continent and consider what seems to be the most natural arrangement. We have thirty millions of people directly before us [the United States], in every way more convenient to us than Canada; they are of the same stock, same feelings as ourselves . . . I do not think that the people of Nova Scotia want annexation to the United States, but why should you drive them against their interests and inclinations into a union with Canada—with which they have no natural means of communication, and no sympathy?

2. **M.I. Wilkins on the Union of British North America**
   (Halifax *British Colonist*, 13 June 1854)

   A union of the B.N.A. Colonies I believe to be indispensable. . . . We may become independent without casting off our allegiance [to Britain]. Colonies are like children. . . . when they arrive at maturity they require different treatment. . . . They have a right to look about the world and set up for themselves. . . . But does it follow from this that all connection between them must cease? By no means. With regard to a federal union I will waste few words. There is an old saying but a very true one that "Union is strength."

3. **Charles Tupper on the need for a new nation**
   (Halifax *Evening Reporter*, 23 January 1862)

   It must be evident to everyone that as we are now situated we are entirely without name or nationality, destitute of all influence and of the means of occupying that position to which we may justly aspire. What is a British [North] American but a man dependent on an Empire which, however glorious, gives him no share or interest in it!

4. **On Charles Tupper**
   (Halifax *Citizen*, 5 November 1864)

   (H)aving got power, Tupper intends to keep it at all hazard, and hence he flies to the Confederation scheme. . . . We have very little faith in Confederation as a practical measure; but we have a good deal of faith in Tupper as a political intriguer. Nothing can serve his turn better just now than to divert public attention from Provincial politics. He would annex this province to Canada, or to Massachusetts, or to the moon, or propose to do so, if by that means he could keep people from talking about his school bill, his retrenchment, his railway duplicity, his tyranny to officials.

5. **On Union**
   (Halifax *British Colonist*, 28 February 1865)

   Union is immediately necessary on account of commercial causes. . . . Union is immediately necessary so as to abolish all differences in currency, and in trade regulations. Each little Province has now its own Government, and even its own postage stamp. Prince Edward Island has one system of currency; Nova Scotia another; New Brunswick another; Newfoundland a fourth; and Canada a fifth. Here is endless confusion and a fruitful source of local jealousy, of sectional alienation, and narrow provincialism. . . . Union is immediately necessary for political reasons. The Nova Scotian is now without a country. He cannot call himself an American. He is not an Englishman. As a Nova Scotian he is nothing.

6. **John A. Macdonald for the Intercolonial Railway**
(A speech in Halifax, 12 September 1864)

(T)his railway must be a national work, and Canada will cheerfully contribute to the utmost extent in order to make that important link, without which no political connection can be complete. What will be the consequence to this city, prosperous as it is, from that communication? Montreal is at this moment competing with New York for the trade of the great West. Build the road and Halifax will become one of the great emporiums of the world. All the great resources of the west will come over the immense railways of Canada to the bosom of your harbour.

Prince Edward Island – Background

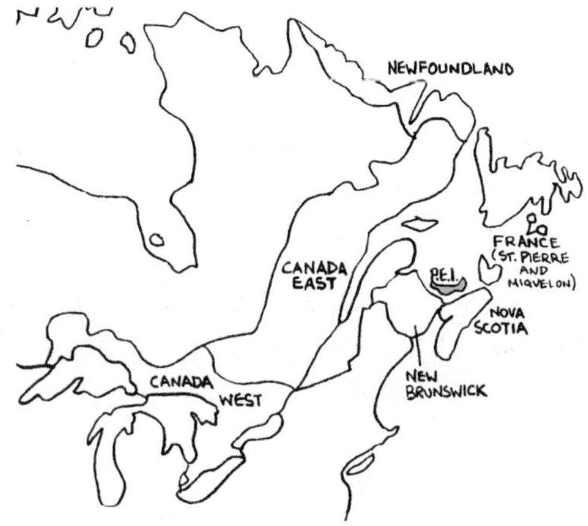

Population (circa 1860): 80,000

Urban centre: Charlottetown (7,000)

Key figures: John Hamilton Gray (1812-1887)
Edward Palmer (1809-1889)
William H. Pope (1825-1879)
Andrew A. Macdonald (1829-1912)
George Coles (1810-1875)
Thomas H. Haviland (1822-1895)
Edward Whelan (1824-1867)

Made up of small, agricultural communities, Prince Edward Island is only one hundred miles long and thirty miles wide. The population is equally divided between Irish Catholics and Scottish Protestants. Largely self-sufficient, many of the residents have never been off the island. During the winter months Prince Edward Island is completely cut off by ice from the mainland. Farming, with the island's rich and fertile lands, is the main source of income. A great many people live on agricultural exports (potatoes, cereal and livestock) to America's New England states. Though a number of residents are fishermen, the United States dominates the island's fishery.

Charlottetown is the colony's urban centre. In fact, no one on the island lives more than a day's journey from the city. With rich church architecture and the impressive Legislature Building, Charlottetown is the colonial capital.

Sixty percent of the farmers on the island rent their land (they are tenant farmers). These islanders rent farms from people who live in Britain (absentee landlords). For over one hundred years, these tenant farmers have been clearing and working the land they do not own and sending rent to their landlords in Britain. Prince Edward Island wants Great Britain to force the landlords to sell their farms at a reasonable price, to the colonists.

After three to four generations as tenant farmers, resentment has grown and left the islanders suspicious of outsiders. To many, Ottawa is as distastefully distant as Imperial London. As an island, a railway link with the other British North American colonies is not an issue. Ultimately, Prince Edward Island is content to remain independent and detached from British North American society.

Though Prince Edward Island does not directly border on the United States and islanders do not fear an American attack, Britain's changing relationship with the colonies is of some concern. Great Britain's reluctance to defend the colonies in the future has made closer ties between the British North American provinces important to security.

Briefing Sheet **12**

Prince Edward Island – Viewpoints

Drawn from P.B. Waite, *Confederation, 1854-1867* (Holt, 1972) and Waite, *Pre-Confederation* (Prentice, 1965).

1. **"Prospects of Confederation in Prince Edward Island"**
   (Charlottetown *Islander*, 6 January 1865)

   The majority of people are under the impression that Confederation would ruin the Island. They have been told that if the Island should be united with the other Provinces, under a Federal government, the people would be heavily taxed—that they would be marched away to the frontiers of Upper Canada to fight for the Canadians; and that the completion of the intercolonial Railroad would bring the produce of Canada to St. John and Halifax, and thus injure the farmers of P.E. Island. Firmly believing that all these evils would come upon them were they to enter the Confederation the people, with but few exceptions, are unanimous in the cry "away with Confederation—we will have nothing to do with it." Confederation, in the terms of the resolutions of the [Quebec] Conference, will, we assume, at the coming Session, be almost unanimously rejected in both Houses of the Legislature of this Island.

2. **On Union**
   (Charlottetown *Examiner*, 22 August 1864)

   Shall we, then, think seriously about a Federal Union. We believe we ought. Great Britain is constantly urging upon our attention a Union of some kind. The only kind of Union we can have is a Federal one. That means little or nothing short of separation from Great Britain. . . . If we make up our minds for an Independent Federation. . . . we must prepare to bid goodbye to old Mother England.

3. **On debt**
   (Charlottetown *Examiner*, 22 October 1864)

   Canada proposes to deal with the Maritime Provinces in the most broad and liberal spirit. She emphatically declares that the burden of her debt shall fall upon Upper and Lower Canada—*and upon Upper and Lower Canada alone.*

4. **Edward Palmer at the Quebec Conference: Fears for the Future of Prince Edward Island**
   (Charlottetown *Monitor*, 15 December 1864)

   I fear our little Island is to be sacrificed [in the Federal Union]. The Canadian ministers or their leaders are clever and ingenious men. They are in this position: their government must stand or fall in the accomplishment or failure of the Union. . . . The paltry proportion of representation we are *now* likely to have in both branches of the [central] Legislature is little more than nominal, and leaves us at the mercy of the other Provinces. . . . In short, I am thoroughly disgusted at the course things have taken here and would be disposed to "sit. . . . and weep" for years, if I thought our Island people would be taken in by the scheme.

Maps of the colonies in 1864

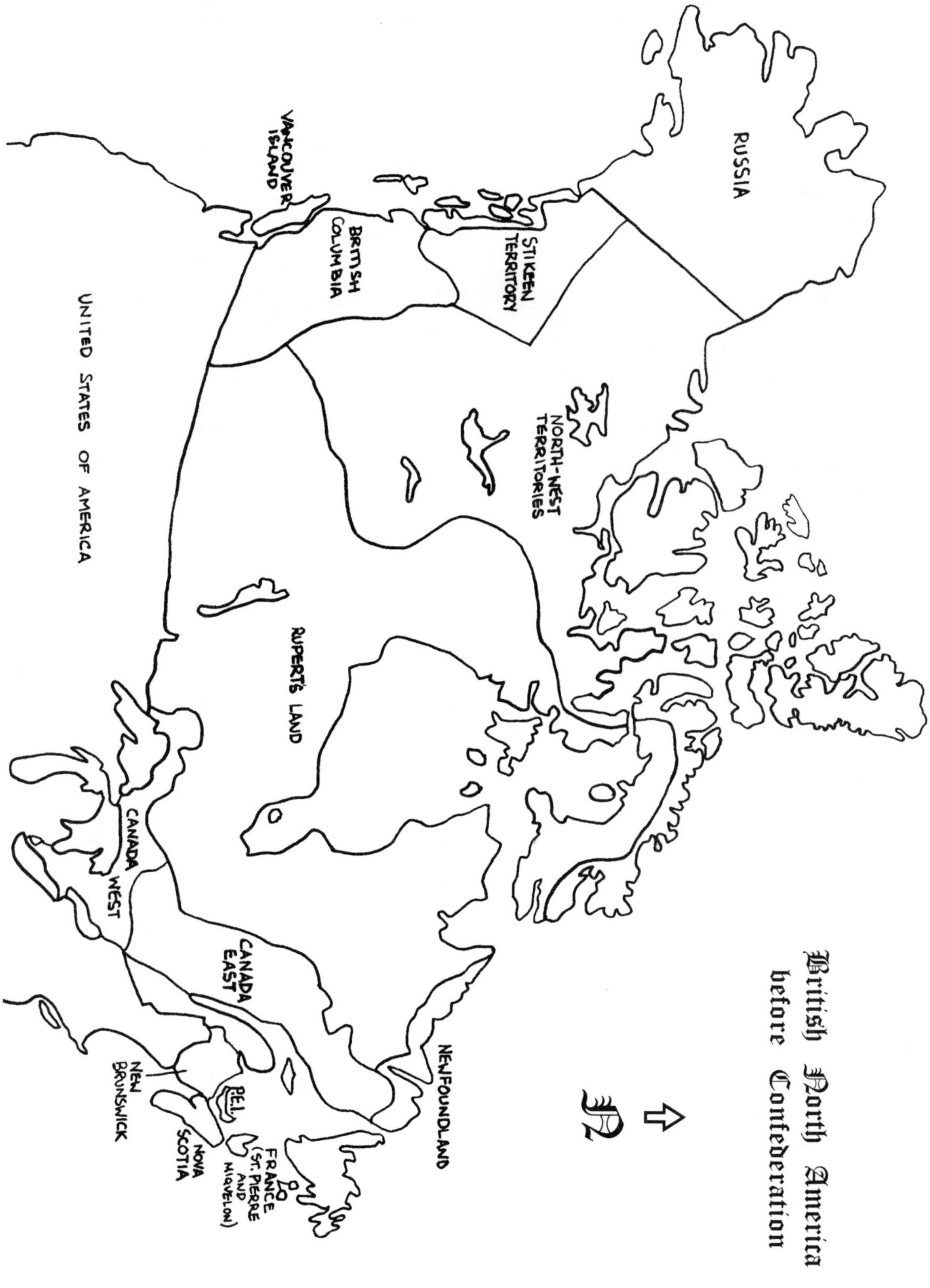

Briefing Sheet
page 1 of 2

British North America before Confederation

14

Daniel Starling, Student
Lord Byng Secondary School

Railways and Canals before Confederation

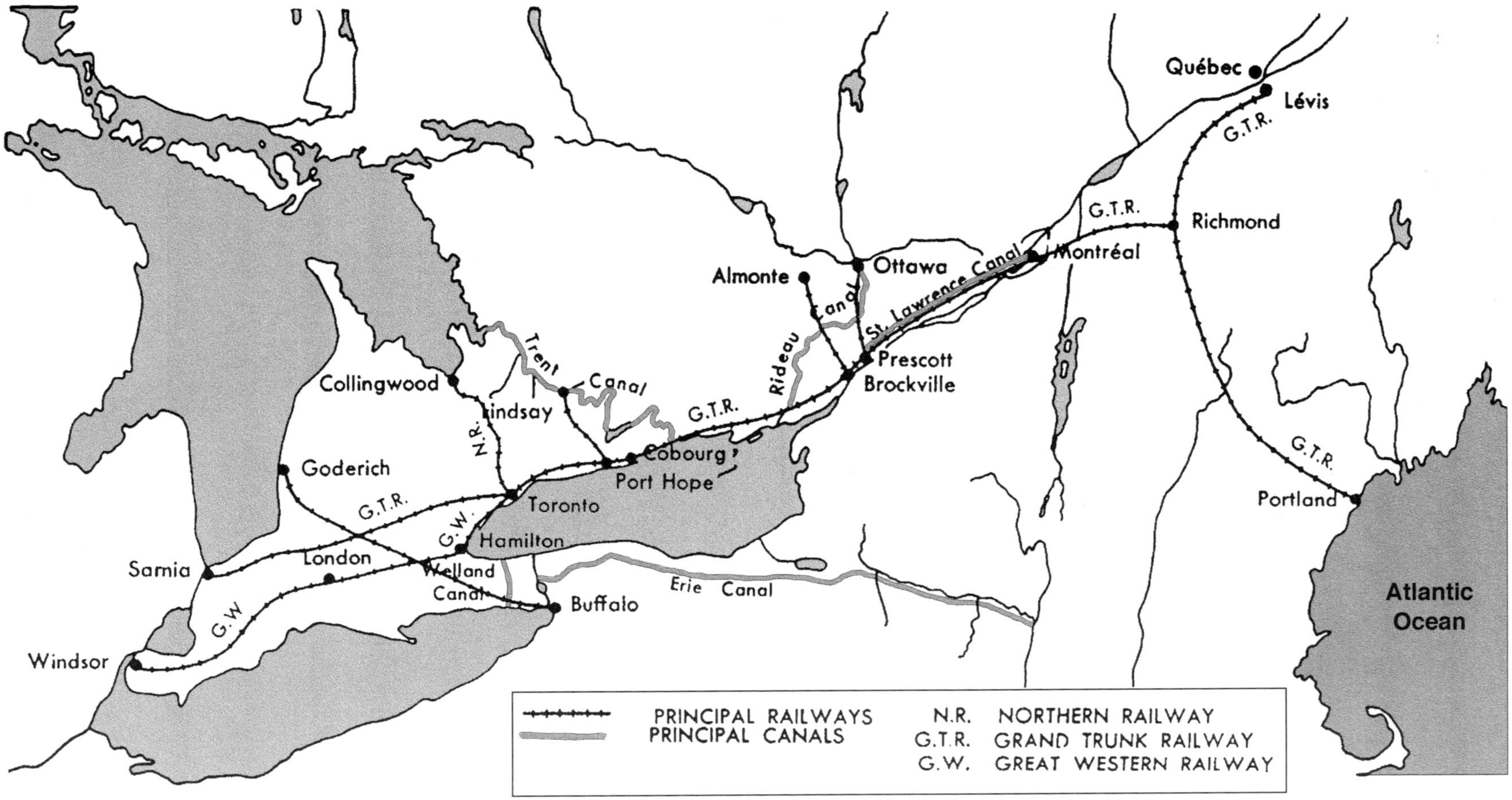

Adapted from: Cornell, Hamelin, Ouellet, Trudel, Canada. *Unity in Diversity*, p. 239.

Annotated bibliography on Confederation

For general background to Confederation the three most useful books are still Donald Creighton, *The Road to Confederation: The Emergence of Canada, 1863-1867*; W.L. Morton, *The Critical Years: The Union of British North America, 1857-1873*; and P.B. Waite, *The Life and Times of Confederation, 1864-1867: Politics, Newspapers, and the Union of British North America*. The Creighton and Morton texts are long narratives, telling the story of Confederation in all of the British North American colonies. Waite's book, however, combines narrative with a generous use of primary sources.

If you are looking for an abundance of **primary sources**, Waite's work is excellent. His *Confederation, 1854-1867* contains a selection of newspaper articles from each of the six colonies. This book is highly readable and engaging: a large portion of the primary sources used in this simulation are taken from it. Other useful books containing a variety of primary sources are J.M. Bliss, *Canadian History in Documents, 1763-1966*; R.C. Brown and M.E. Prang, *Confederation to 1949*; H.H. Herstein, L.J. Hughes and R.C. Kirbyson, *Challenge and Survival: The History of Canada*; J.H. Stewart Reid, Kenneth McNaught and Harry S. Crowe, *A Source-Book of Canadian History*; P.B. Waite, *The Confederation Debates in the Province of Canada, 1865*; and also Waite, *Pre-Confederation*.

For more **background on specific colonies**, the following secondary sources are recommended: Francis W.P. Bolger, *Prince Edward Island and Confederation, 1863-1873*; W.S. MacNutt, *New Brunswick: A History, 1784-1867*; and Kenneth G. Pryke, *Nova Scotia and Confederation, 1864-74*.

General Canadian history surveys which also cover Confederation are Vivien Bowers and Stan Garrod, *Our Land: Building the West* ; Penney Clark and Roberta McKay, *Canada Revisited: A Social and Political History of Canada to 1911*; Donald Creighton, *Dominion of the North: A History of Canada*; R. Douglas Francis, Richard Jones and Donald B. Smith, *Origins: Canadian History to Confederation*; J.L. Granatstein, et al. *Nation: Canada Since Confederation*; H.H. Herstein, *Challenge and Survival*; and Diane Eaton and Garfield Newman, *Canada: A Nation Unfolding*.

The following books are **anthologies of secondary sources**, containing shorter, more academic pieces, but suitable for advanced reading: J.M. Bumsted, *Interpreting Canada's Past: Volume One, Before Confederation*; R. Douglas Francis and Donald B. Smith, *Readings in Canadian History: Pre-Confederation*; Francis and Smith, *Readings in Canadian History: Post-Confederation*; and G.A. Rawlyk, *Historical Essays on the Atlantic Provinces*. A more comprehensive list of readings may be found in Francis, Jones and Smith's *Origins* (pages 395-396).

References

Bliss, J.M. (ed.). (1966). *Canadian History in Documents, 1763-1966.* Toronto: Ryerson.

Bolger, Francis W.P. (1964). *Prince Edward Island and Confederation, 1863-1873.* Charlottetown, PEI: St. Dunstan's University Press.

Bowers, Vivien & Garrod, Stan (1987).*Our Land: Building the West*. Toronto: Gage.

Brown, R.C., & Prang, M.E. (eds.). (1966). *Confederation to 1949* (Volume 3 of *Canadian Historical Documents* Series). Scarborough, ON: Prentice.

Bumsted, J.M. (ed.) (1986). *Interpreting Canada's Past: Volume One, Before Confederation.* Toronto: Oxford University Press.

Clark, Penney & McKay, Roberta. (1992). *Canada Revisited: A Social and Political History of Canada to 1911.* Edmonton: Arnold.

Creighton, Donald. (1957). *Dominion of the North: A History of Canada.* Toronto: Macmillan.

Creighton, Donald. (1964). *The Road to Confederation: The Emergence of Canada, 1863-1867.* Toronto: Macmillan.

Eaton, Diane & Newman, Garfield (1994). *Canada: A Nation Unfolding.* Toronto: McGraw-Hill.

Francis, R. Douglas, Jones, Richard & Smith, Donald B. (1988). *Origins: Canadian History to Confederation.* Toronto: Holt.

Francis, R. Douglas & Smith, Donald B. (1986). *Readings in Canadian History: Pre-Confederation.* (2nd edition). Toronto: Holt.

Francis, R. Douglas & Smith, Donald B. (1986). *Readings in Canadian History: Post-Confederation.* (2nd edition). Toronto: Holt.

Granatstein, J.L., et al. (1990). *Nation: Canada Since Confederation.* (3rd ed). Toronto: McGraw.

Herstein, H.H., Hughes, L.J. & Kirbyson, R.C. (1970). *Challenge and Survival: The History of Canada.* Scarborough: Prentice.

MacNutt, W.S. (1963). *New Brunswick: A History, 1784-1867.* Toronto: Macmillan.

Morton, W.L. (1964). *The Critical Years: The Union of British North America, 1857-1873.* Toronto: McClelland.

Pryke, Kenneth G. (1979). *Nova Scotia and Confederation, 1864-74.* Toronto: University of Toronto Press.

Rawlyk, G.A. (1967). *Historical Essays on the Atlantic Provinces.* Toronto: McClelland.

Reid, J.H. Stewart, McNaught, Kenneth & Crowe, Harry S. (1959). *A Source-Book of Canadian History.* Toronto: Longmans.

Waite, P.B. (1962).*The Life and Times of Confederation, 1864-1867: Politics, Newspapers, and the Union of British North America.* Toronto: University of Toronto Press.

Waite, P.B. (ed.). (1963). *The Confederation Debates in the Province of Canada, 1865.* Toronto: McClelland.

Waite, P.B. (ed.). 1965. *Pre-Confederation.* (Volume 2 of *Canadian Historical Documents* Series). Scarborough, ON: Prentice.

Waite, P.B. (1972). *Confederation, 1854-1867.* Toronto: Holt.

Wallace, Stuart W. (1985). *The Macmillan Dictionary of Canadian Biography* (4th edition). Toronto: Macmillan.

Father of Confederation profile

<table>
<tr><td colspan="2">Name of delegate</td><td rowspan="5">photocopy photo
and place here</td></tr>
<tr><td>Colony represented</td><td>Conferences attended</td></tr>
<tr><td>Date of birth</td><td>Place of birth</td></tr>
<tr><td>Nationality at birth</td><td>Date of death</td></tr>
<tr><td>Age at Confederation (1867)</td><td>Gender</td></tr>
<tr><td>First language</td><td>Political affiliation</td><td>Religion</td></tr>
<tr><td colspan="3">Profession or career (outside of politics)</td></tr>
<tr><td colspan="3">Brief summary of person's role in Confederation</td></tr>
<tr><td colspan="3">List of other achievements or notable events</td></tr>
</table>

Student Activity 16

Colonial perspective

Colony: ___________________________

Major economic activities
Key groups in colony

In the left-hand column, identify the main needs and concerns of your colony and, in the right-hand column, explain how joining in Confederation may help to solve or to worsen each need / concern.

Main needs/concerns	Implications for Confederation

Student Activity *page 1 of 2* **17**

Potential advantages of Confederation	Potential disadvantages of Confederation
political	
social/cultural	
economic	
military	

Preliminary proposal

Colony: ________________________

Conditions for joining Confederation	Justification

Use another sheet if necessary

Student Activity 18

Final proposal

Colony: ______________________________

Conditions carried over from preliminary proposal	Justification for continued insistence

Modified, deleted or new conditions	Justification for change

Use another sheet if necessary

Student Activity **19**

Decision on Confederation Bill

Colony: ______________________________

Issues resolved to the colony's satisfaction	Anticipated benefits

Unresolved issues	Anticipated negative consequences

Our delegation has decided to ❑ accept the Confederation Bill
❑ reject

Key reasons for our decision

Student Activity *page 2 of 2* **20**

Reflecting on the experience

1. Which colonies bargained from a position of strength? Why?

2. Which colonies bargained from a position of weakness? Why?

3. Describe your colony's efforts and degree of success at negotiating its demands.

4. List and explain three *effective* and three *ineffective* negotiating techniques that you observed during the Ottawa Conference.

effective techniques	ineffective techniques
•	•
•	•
•	•

5. How might Canada be different today if the British North America Act signed in this simulation actually became law?

6. What did you enjoy about the Ottawa Conference?

7. What did you dislike about the Ottawa Conference?

8. If this simulation were to be repeated, what might you personally have done differently to make the experience more successful?

Conference agenda

Orientation

- Introduction to the simulation
- Organization of colonial delegations

Preliminary Proposals

- Briefing on colonial issues
- Independent research
- Preparation of preliminary demands
- Presentation of preliminary proposals

Final Proposals

- Study of other colonies' proposals
- Preparation of final demands
- Presentation of final proposal

Voting on the Bill

22

Support Material

Conference materials

Representative: ________________________________

1864 Ottawa Conference on Confederation

23

Support Material

Representatives' name tags

George-Etienne Carter
Canada East

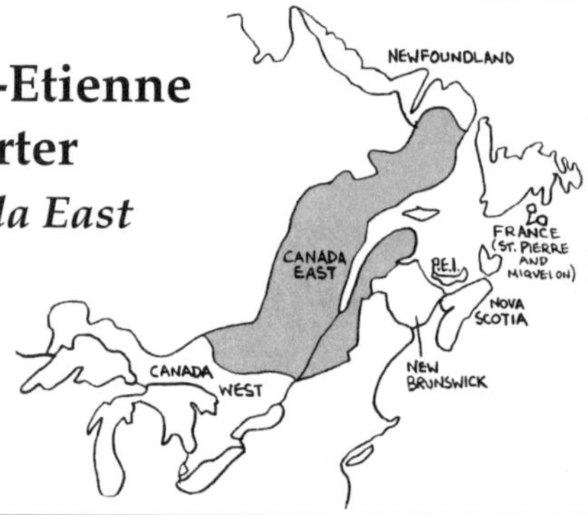

Jean-Charles Chapais
Canada East

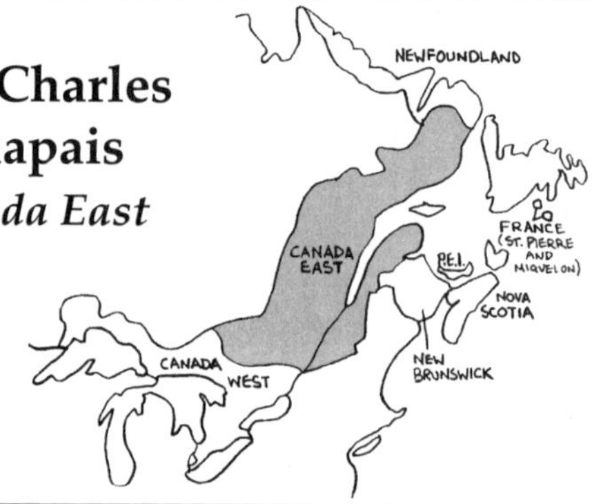

Alexander T. Galt
Canada East

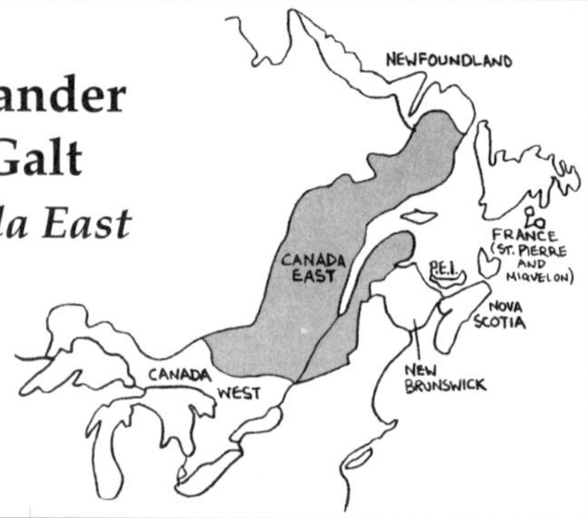

Hector L. Langevin
Canada East

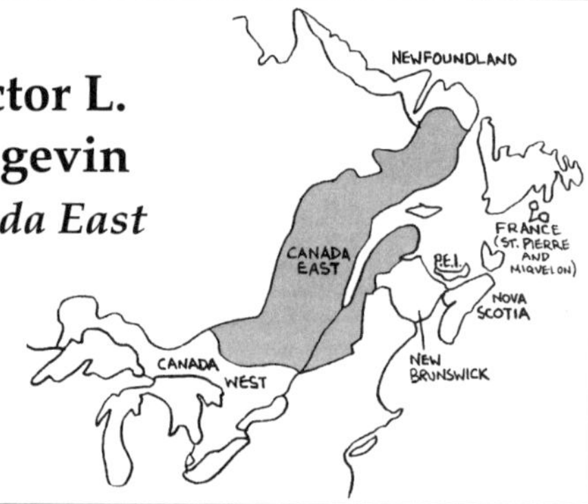

Thomas D'Arcy McGee
Canada East

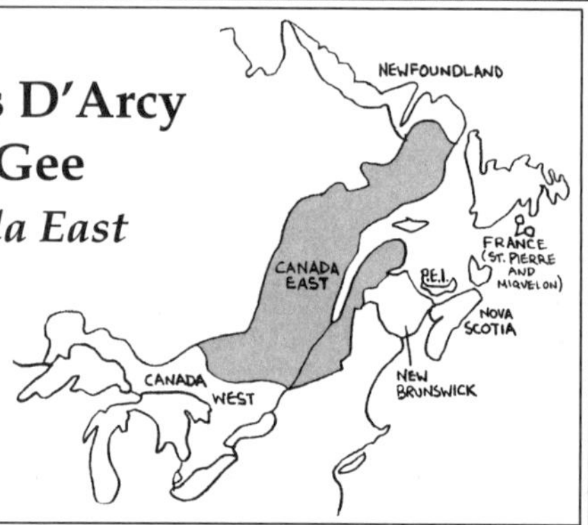

Sir Etienne-Paschal Taché
Canada East

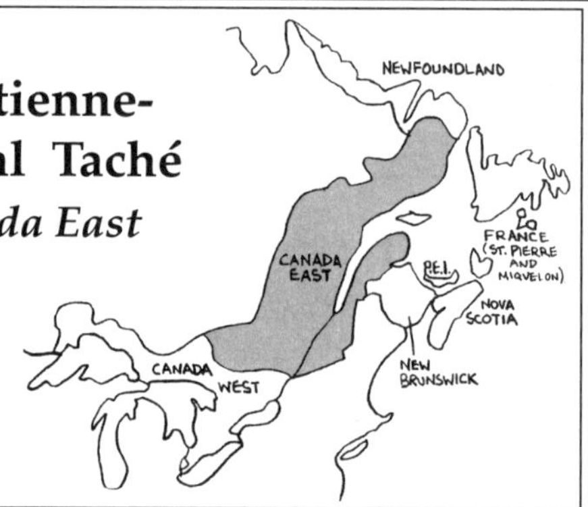

24

Support Material
page 1 of 6

George Brown
Canada West

Alexander Campbell
Canada West

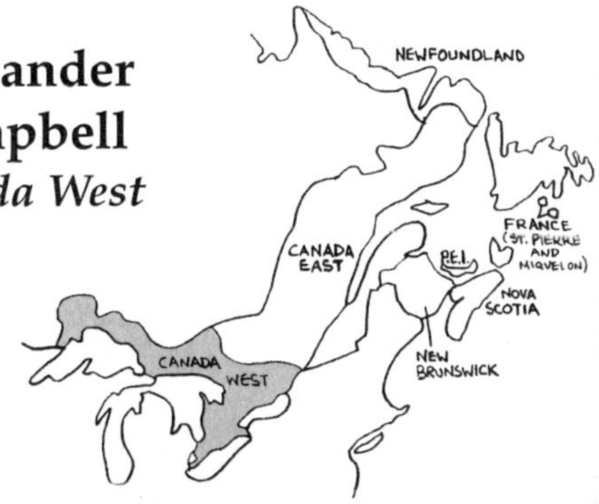

James Cockburn
Canada West

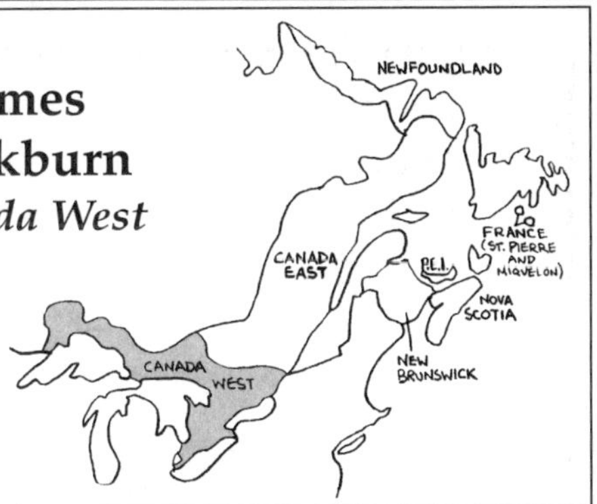

William P. Howland
Canada West

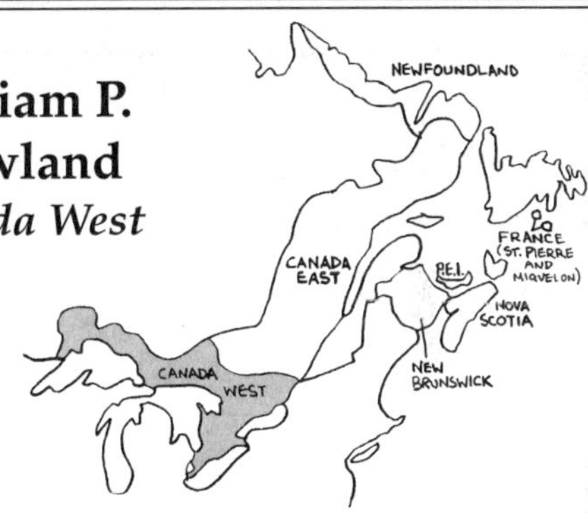

John A. Macdonald
Canada West

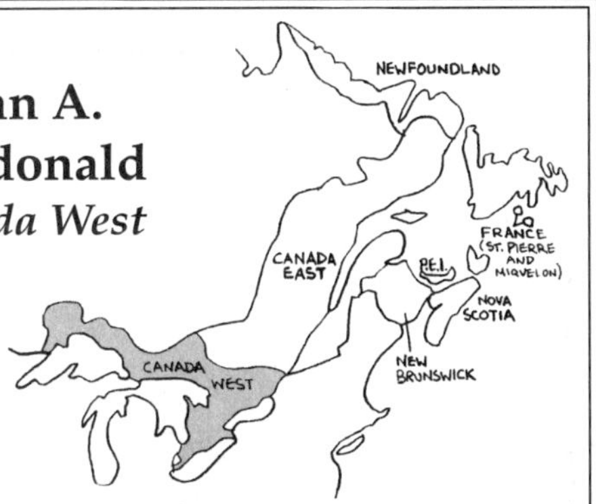

William Macdougall
Canada West

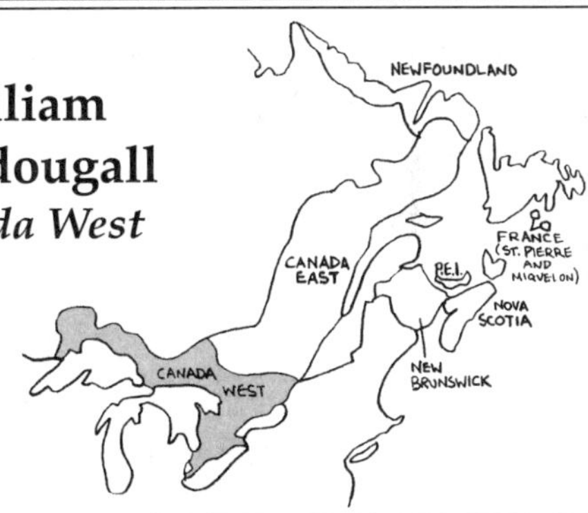

Oliver Mowat
Canada West

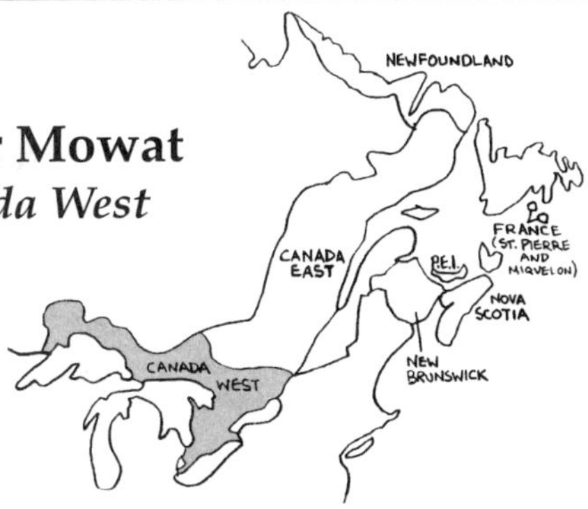

Edward B. Chandler
New Brunswick

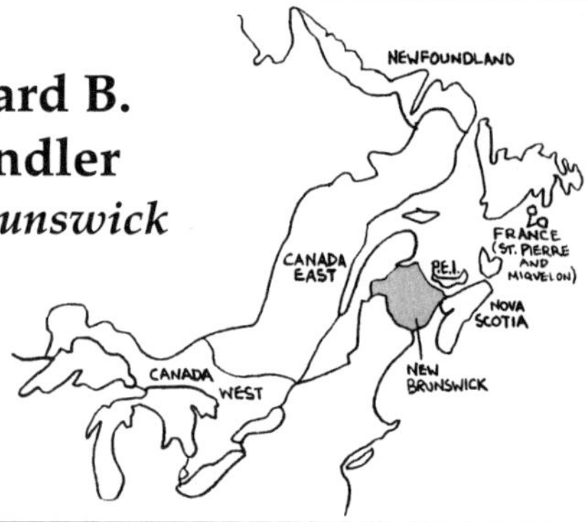

Charles Fisher
New Brunswick

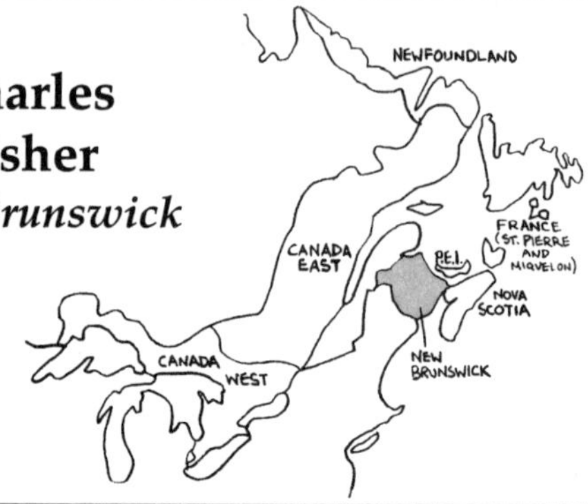

John Hamilton Gray
New Brunswick

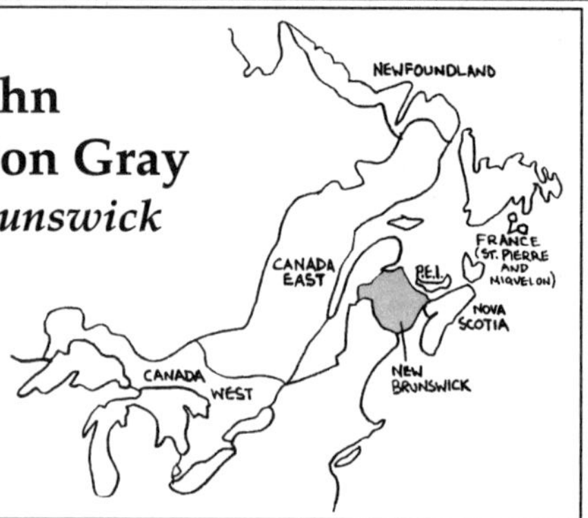

John M. Johnson
New Brunswick

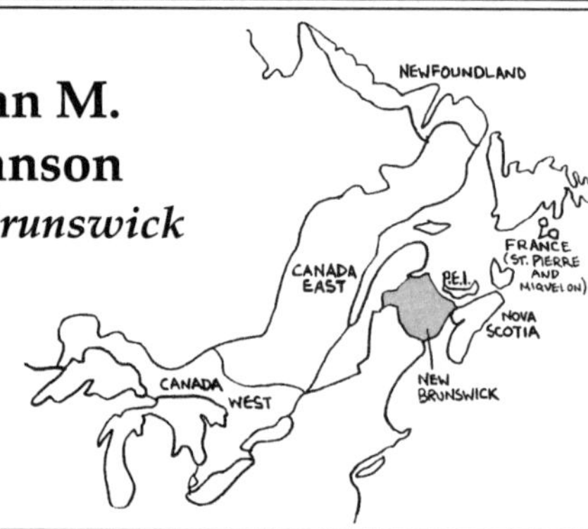

William H. Steeves
New Brunswick

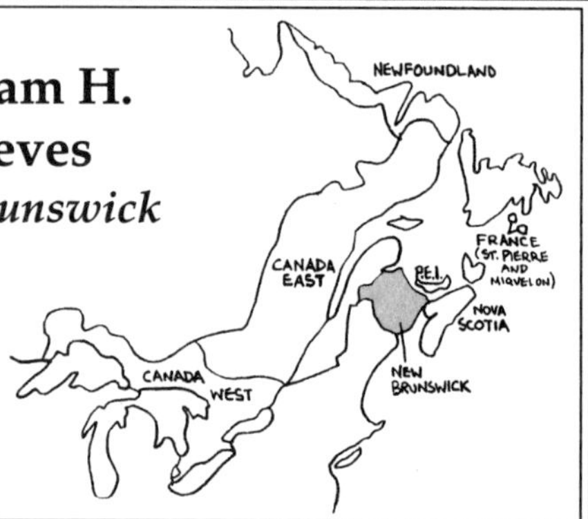

Samuel Leonard Tilley
New Brunswick

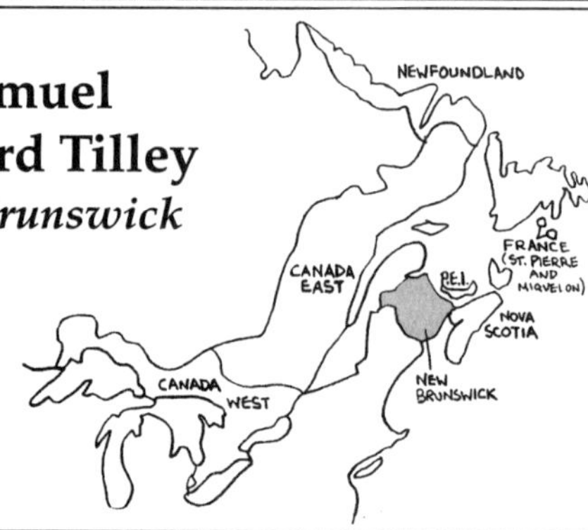

Frederick B.T. Carter
Newfoundland
(Re-assigned from New Brunswick)

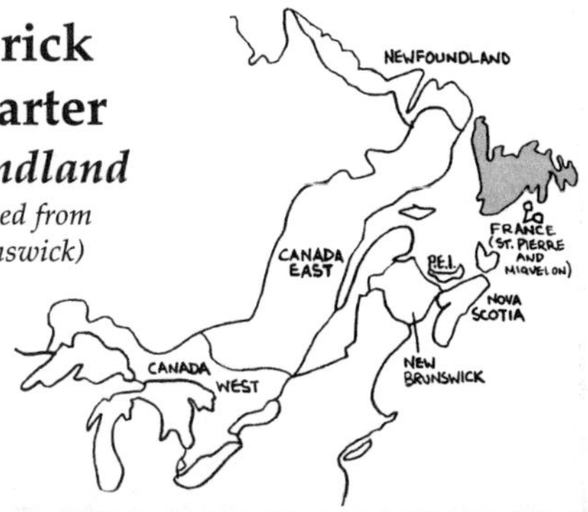

Peter Mitchell
Newfoundland

Ambrose Shea
Newfoundland

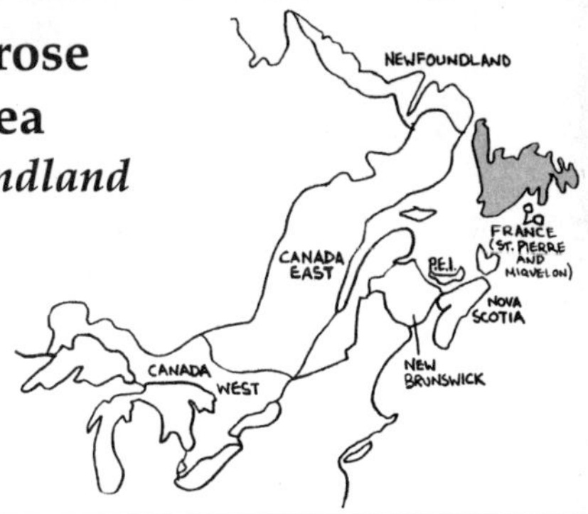

Edward Whelan
Newfoundland
(Re-assigned from P.E.I.)

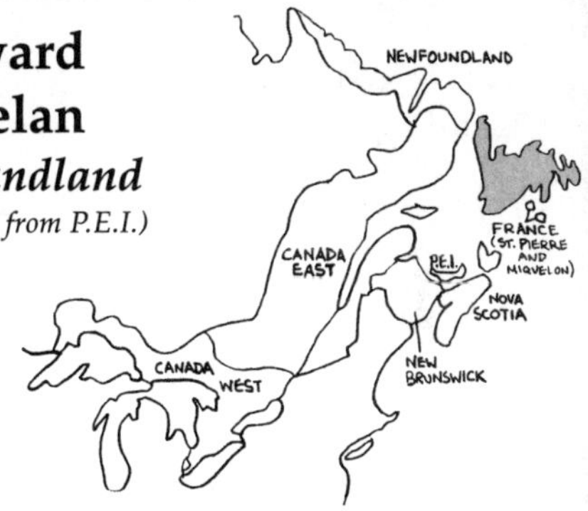

R.D. Wilmot
Newfoundland
(Re-assigned from New Brunswick)

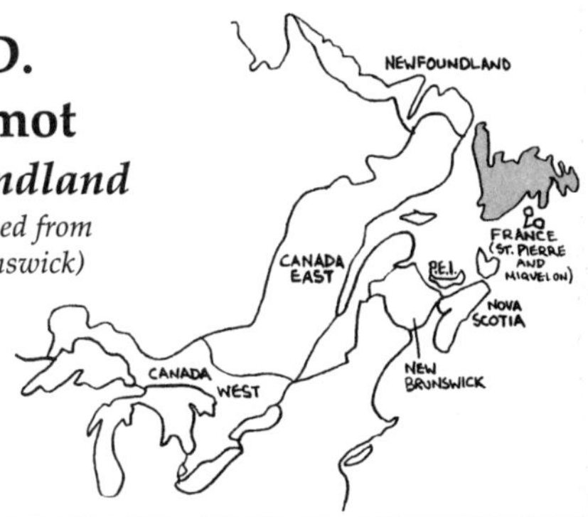

24

Support Material
page 4 of 6

Adams G. Archibald
Nova Scotia

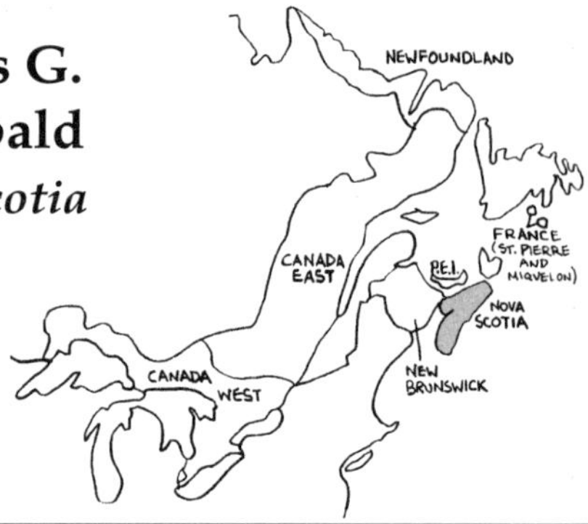

Robert B. Dickey
Nova Scotia

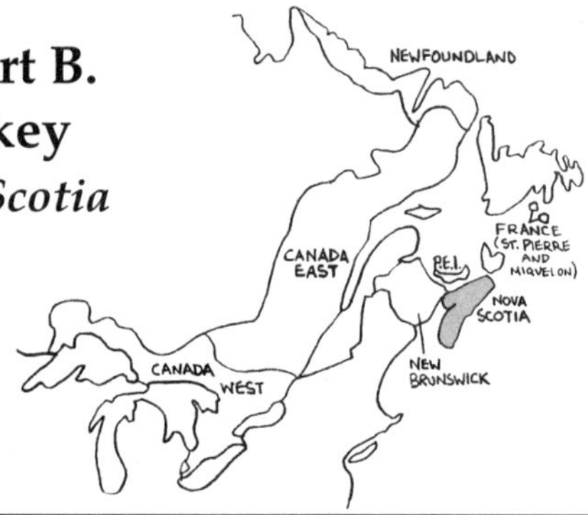

William A. Henry
Nova Scotia

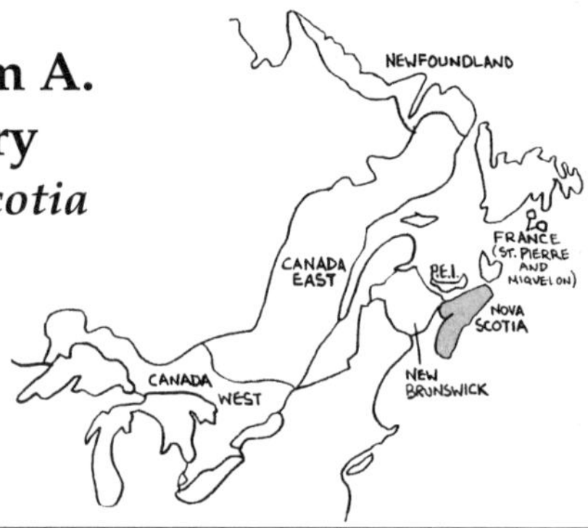

Jonathon McCully
Nova Scotia

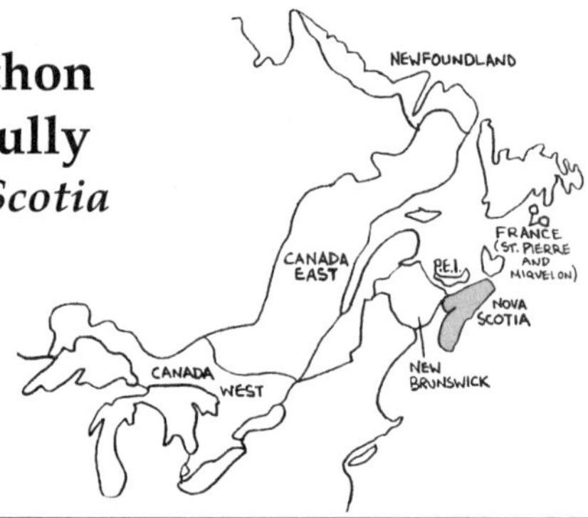

John W. Ritchie
Nova Scotia

Charles Tupper
Nova Scotia

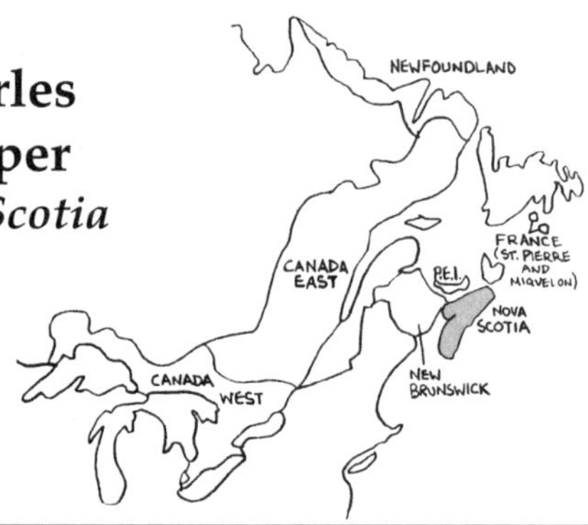

George H. Coles
Prince Edward Island

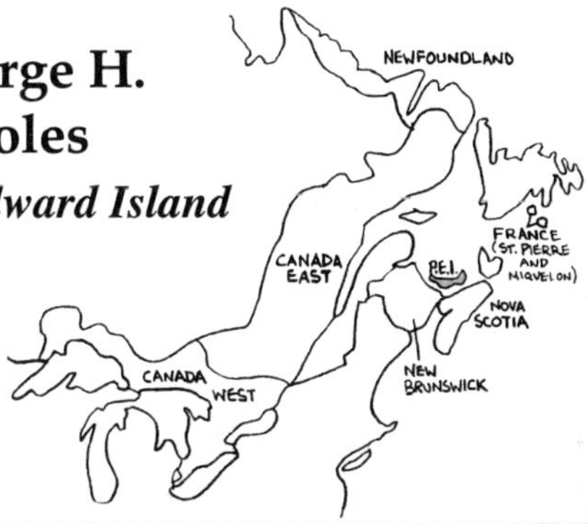

John Hamilton Gray
Prince Edward Island

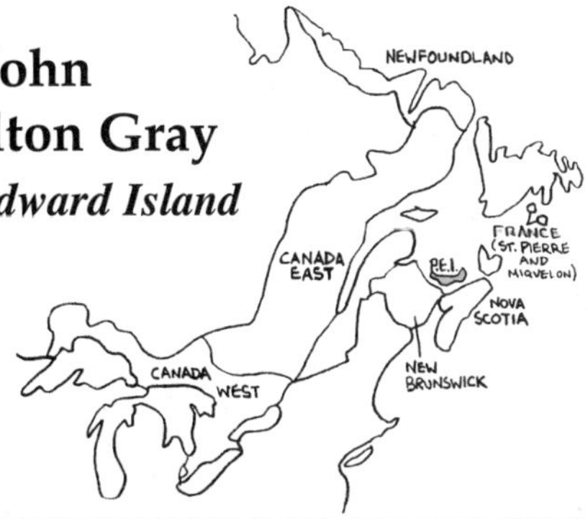

Thomas Heath Haviland
Prince Edward Island

Andrew A. Macdonald
Prince Edward Island

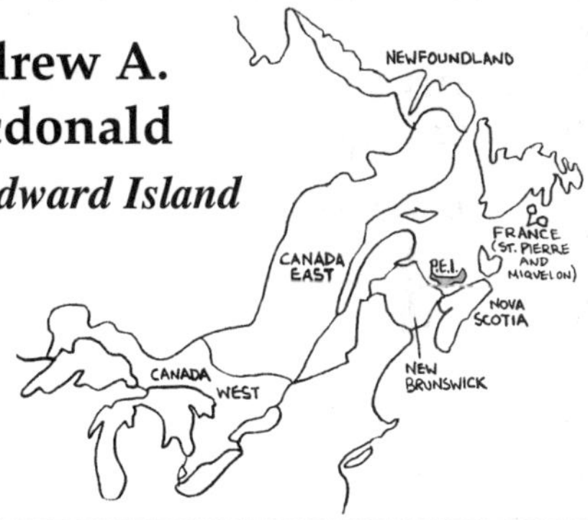

Edward Palmer
Prince Edward Island

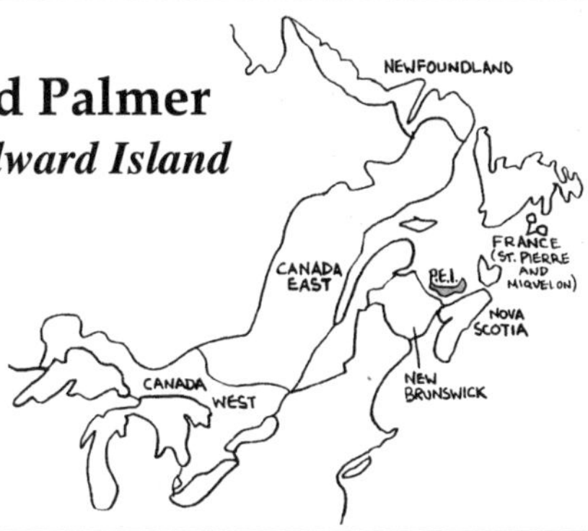

William H. Pope
Prince Edward Island

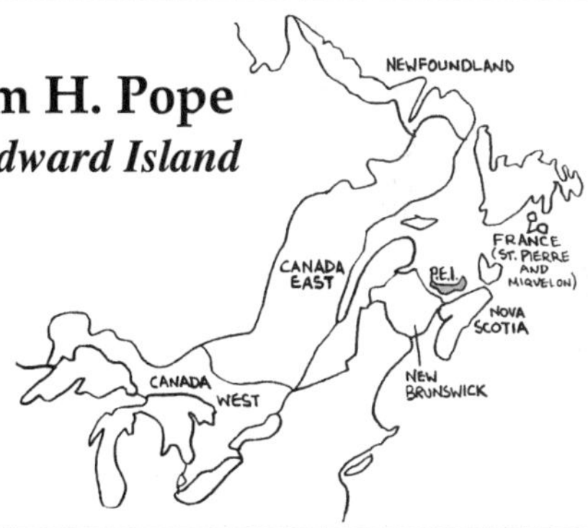

Confederation Bill, 1867

This proposed legislation, to be voted on by the six member colonies of British North America, was written by Her Majesty the Queen's Parliament of the United Kingdom of Great Britain and Ireland; it is based on the recommendations presented by the colonies, at the conference at Ottawa, Canada West, in 1864.

Terms of the Federal Union of the Colonies of British North America:

25

Support Material

page 1 of 2

Signatories to Confederation

On behalf of Canada East

On behalf of Canada West

On behalf of New Brunswick

On behalf of Newfoundland

On behalf of Nova Scotia

On behalf of Prince Edward Island

On behalf of Her Majesty Queen Victoria

Date

25

Support Material

page 1 of 2

Knowledge of colonial perspectives

Your colony: ________________________________

Explain two major concerns with Confederation for your colony and why each is a concern. **/ 4**

Concern	Explanation for concern

Explain three major benefits to your colony in joining Confederation and why each is desirable. **/ 6**

Benefit	Explanation for benefit

Assessment
page 1 of 2

26

Identify the five other colonies represented at this Conference and describe a major concern with and a major benefit from Confederation. **/ 15**

Name of colony	Concern with Confederation	Benefit from Confederation

TOTAL **/ 25**

Assessment *page 2 of 2*

26

Preliminary proposals

Use this rubric to assess Blackline Master #18 and the accompanying presentation.

Indicate which delegation is being assessed by checking the appropriate box.

- ❑ Canada East
- ❑ Canada West
- ❑ New Brunswick
- ❑ Newfoundland
- ❑ Nova Scotia
- ❑ Prince Edward Island

Contents of proposal / 20

	Definitely not evident	**Somewhat evident**	**Largely evident**	**Fully evident**
Requirements are **clear and specific**	1	2	3	4
Requirements **address key needs** of the colony	1	2	3	4
Requirements are **historically plausible**	1	2	3	4
Justifications are **realistic and convincing**	2	4	6	8

Manner of presentation / 20

	Definitely not evident	**Somewhat evident**	**Largely evident**	**Fully evident**
Audible and clearly articulated presentation	1	2	3	4
Clear and effective **structure** to presentation	1	2	3	4
Appropriate and effective **use of visual aids**	1	2	3	4
Full participation in presentation by entire delegation	2	4	6	8

TOTAL / 40

Comments:

Final proposals

Use this rubric to assess Blackline Master #19 and the accompanying presentation.

Indicate which delegation is being assessed by checking the appropriate box.

- ❑ Canada East
- ❑ Canada West
- ❑ New Brunswick
- ❑ Newfoundland
- ❑ Nova Scotia
- ❑ Prince Edward Island

Contents of proposal / 20

	Definitely not evident	Somewhat evident	Largely evident	Fully evident
Requirements are **clear and specific**	1	2	3	4
Requirements **address key needs** of the colony	1	2	3	4
Requirements are **historically plausible**	1	2	3	4
Justifications are **realistic and convincing**	2	4	6	8

Manner of presentation / 20

	Definitely not evident	Somewhat evident	Largely evident	Fully evident
Audible and clearly articulated presentation	1	2	3	4
Clear and effective **structure** to presentation	1	2	3	4
Appropriate and effective **use of visual aids**	1	2	3	4
Full participation in presentation by entire delegation	2	4	6	8

TOTAL / 40

Comments:

Final decisions

Use this rubric to assess Blackline Master #20.

Indicate which delegation is being assessed by checking the appropriate box.

- ❑ Canada East
- ❑ Canada West
- ❑ New Brunswick
- ❑ Newfoundland
- ❑ Nova Scotia
- ❑ Prince Edward Island

	Weak	Satisfactory	Very good
Identification of of resolved issues	Fails to correctly identify any key issues resolved by the proposed bill **1**	Identifies three resolved issues but either misses major ones or is somewhat confused about them **3**	Clearly identifies three or more issues resolved by the proposed bill, and no key issues are missed **5**
Anticipated benefits	Lacks plausible benefits arising from the proposed bill **1**	Identifies benefits arising arising from three issues, but either misses major benefits or they are somewhat unrealistic **3**	Clearly explains plausible benefits for at least three issues, and no major benefits are overlooked **5**
Identification of unresolved issues	Lacks correctly identified key issues unresolved by the proposed bill **1**	Identifies three unresolved issues but either misses major ones or is somewhat confused about them **3**	Clearly identifies at least three issues unresolved by the proposed bill and includes all key issues **5**
Anticipated negative consequences	Lacks plausible negative consequences arising from the proposed bill **1**	Identifies negative consequences arising from three unresolved issues, but either misses major negative consequences or they are somewhat unrealistic **3**	Clearly explains plausible negative consequences for three or more unresolved issues, and includes major negative consequences **5**
Justification for decision	Either offers no justification or reasons are largely trivial or irrelevant to the decision **1**	Offers three relevant reasons, which may be vaguely stated or ignored **3**	Offers three or more clearly articulated, plausible reasons which address the important issues for the colony **5**

TOTAL /25

Comments:

Assessment

29

Individual contributions

	Very poor	Weak	Competent	Accomplished	Outstanding
Contribution to final product	Individual contributed no useful ideas and did little or no work **0**	Individual contributed very few productive ideas and did much less than his/her equal share of the work **3**	Individual contributed productive ideas and did his/her equal share of the work **6**	Individual regularly contributed productive and thoughtful ideas and was generous in the amount of work done for the project **9**	Individual regularly provided extremely productive and thoughtful ideas that profoundly shaped the project OR did more than his or her share of the work **12**
Contribution to group dynamic	Individual was a major and constant impediment to the friendly and efficient operation of the group **0**	Individual often seriously impeded the friendly and efficient operation of the group **2**	Individual may only occasionally have impeded the efficient operation of the group, and participated most of the time but was not a key factor in shaping the group dynamic **4**	Individual almost never impeded the friendly and efficient operation of the group and most of the time was a positive force in shaping the group dynamic **6**	Individual regularly made significantly empowering or nurturing contributions to the friendly and efficient operation of the group **8**

Assess *yourself* and *each member* of your group on the basis of this rubric by deciding which description most accurately reflects that person's actions within the group. For each person, assess the contribution to the final product and to the group dynamic by writing the number in the space provided on the next page. Assign partial marks if a person's contribution falls between the scale. Explain the reasoning or identify a typical behaviour that is supporting evidence for your assessment.

Do not show or discuss your assessment with anyone else. You will be assessed on how carefully and fairly you assess your peers.

	Contribution to final product		Contribution to group dynamic		
	SCORE	EXPLANATION	SCORE	EXPLANATION	TOTAL
________ Your name	/12		/8		/20
________ Group member's name	/12		/8		/20
________ Group member's name	/12		/8		/20
________ Group member's name	/12		/8		/20
________ Group member's name	/12		/8		/20
________ Group member's name	/12		/8		/20

Assessing student assessments

Use this rubric to assess how well students have assessed their own individual contributions and those of their peers as required in Blackline Master #30. Award intermediate marks for assessments falling between the descriptors.

Rating	Description
Excellent **10**	Clear sense that student has been **conscientious and fair** in assigning ratings. Ratings are supported with **appropriate and specific evidence. All ratings appear warranted** by the evidence provided by the student and are **highly consistent** with reliable assessments by other students and with the teacher's own observations.
Good **7**	A sense that student has **attempted to be conscientious and fair** in assigning ratings. Most ratings are supported with **appropriate evidence.** Virtually all ratings appear **warranted** by the evidence provided by the student and are **reasonably consistent** with reliable assessments by other students and with the teacher's own observations.
Mediocre **4**	Some suspicion that student has been **neither particularly conscientious nor fair** in assigning ratings. Ratings are supported with **vague or unconvincing evidence.** Some **ratings differ significantly** from reliable assessments by other students and from the teacher's own observations.
Very poor **1**	Strong suspicion that student has been **hasty** or **unfair** in assigning ratings. Ratings are supported with **little or no evidence**. The **ratings differ significantly** from reliable assessments by other students and from the teacher's own observations.

TOTAL **/ 10**

Comments:

Assessment

31

Tool Enhancement Supplement

We should not assume that students will be effective at putting their heads together to think in concert with others. Many students may need to be taught how to contribute to and benefit from participation in a community of thinkers. The purpose of this section is to identify a range of tools and describe 13 activities for helping students cooperatively think critically about four tasks they will face during the simulation:

- contributing to the group
- developing arguments
- presenting ideas
- negotiating agreement.

	Background knowledge	Criteria for judgment
Contributing to the group *Taking an active, effective role in small group activities*	• knowledge of the range of work habits and personalities of group members • knowledge of how to respond in various group situations to meet individuals' needs	• *respectful*: is sensitive to the feelings and opinions of everyone in the group • *fair*: does one's fair share of the duties of the group and does not dominate the group • *on task:* stays focussed on the task at hand • *collaborative*: contributes in appropriate ways to assist in overall group success
Developing arguments *Generating and framing effective arguments*	• knowledge of the topic under consideration • knowledge of one's own shortcomings in analyzing issues	• *clear and elaborated*: conclusions and supporting reasons are unambiguously and adequately articulated • *plausible*: the statements are reasonable given the evidence • *relevant and substantiated*: arguments are supported with relevant reasons and evidence that are sufficient to establish the claim • *well considered*: the significant consequences for all who may be affected and alternative interpretations and perspectives are carefully considered
Presenting ideas *Preparing and delivering effective presentations*	• knowledge of the topic of the presentation • knowledge of audience needs and how to respond to different audiences • understanding of the various types of oral presentations (e.g., formal lecture, interactive-presentation, question-and-answer) and visual presentations (e.g., posters, murals, overheads) • knowledge that presentations serve different purposes (e.g., inform, persuade, critique, entertain)	• *clear*: comments and visuals are clear to everyone • *parsimonious*: takes no more time than is necessary to get message across • *engaging*: presented in a manner that involves audience and invokes interest • *worthwhile*: content of presentation is sound and of value to the audience • *appropriate*: language and medium of expression is suitable to the message and occasion
Negotiating agreement *Preparing for and taking part in effective negotiations*	• knowledge of the goals and difficulties of all parties in the negotiations • knowledge of the personalities and habits of all individuals in the negotiations • knowledge that individuals may see things in significantly different ways	• *reflective of interests*: solution reflects a broad consensus of the group, not simply the interests of the most forceful members • *feasible*: solution is realistic, likely to achieve its goal • *sustainable*: solution is sustainable over the long term

If students are to perform these tasks successfully they require a diverse set of intellectual tools. Although it would not be advisable to attempt to teach every tool listed in the double-paged chart below, being aware of the range of tools that students might need may help in selecting the most appropriate tools to introduce. The lists of tools are also of value for assessment purposes: they provide specific features to look for when evaluating students.

Critical thinking vocabulary	Thinking strategies	Habits of mind
• cooperation and collaboration • leadership • personal responsibility	• *group management strategies* (e.g., taking turns, assigning cooperative roles, keeping a speaker's list) • *interpersonal strategies* (e.g., praises efforts of others, makes eye contact, uses first names, refers to points made by others) • *planning strategies* (e.g., dividing up tasks among group members, creating a work schedule, monitoring completion of tasks)	• *empathy*: an inclination to feel for others in the group • *forthcoming*: a willingness to express personal views • *initiative*: a willingness to take on a role when a need arises (e.g., provide leadership, volunteer to help, offer encouragement) • *self-monitoring:* attention to effects of one's words and actions on the group
• point of view (or perspective) • bias • argument and counter-argument • pro and con • justification versus rationalization • evidence • fact statements and value statements • premise and conclusion • generalization and over-generalization	• looks for pro and con arguments, and counter-arguments • adopts a role to explore and understand other perspectives • employs a systematic issue analysis model: 1. define the issue and explain why it is important 2. research and explain major pro and con arguments 3. evaluate the comparative strength of competing reasons 4. formulate a defensible position that considers competing interests 5. offer counter-arguments in defense of position	• *open-minded*: open to views other than one's own—especially those that are counter to one's own position • *fair-minded*: judges ideas on their merits and not simply to enforce personal interests and bias • *inquiring or critically minded:* does not take everything at face value—is inclined to question when warranted • *independent minded*: resists the pressures to adopt and espouse opinions merely because they are popular • *intellectual work ethic:* committed to thinking thoroughly and carefully, despite the mental effort required
	• uses thoughtful *presentation preparation* strategies (e.g., develops outlines to identify and sequence key ideas; makes frequent use of illustrative examples; rehearses presentation; prepares drafts of oral comments and mock-ups of visual displays for editing) • uses appropriate oral *presentation delivery* strategies (e.g., speaks from prepared notes; uses graphics to summarize information; checks posture and audibility; connects remarks to earlier speaker's comments or to topics familiar to the audience; sets a clear context for the audience; closes with summary or concluding comment)	• *flexible*: willing to adjust prepared presentation as need arises • *pride in work*: cares about preparing quality work—appreciating the value of drafting and rehearsing • *sense of audience*: is inclined to listen to and watch audience for clues
• point of view • unanimous • consensus • minority position • compromise • ends (goals) and means (strategies)	• bargaining strategies (e.g., formulates opening position, identifies bottom-line position) • active listening strategies (e.g., summarizes and repeats) • conflict resolution strategies (e.g., takes turns, looks for underlying interests)	• *open-minded*: open to views other than one's own—especially those that are counter to one' own position • *fair-minded*: judges ideas on their merits and not simply to enforce personal interests and bias • *patient:* is willing to allow others the time they need to work things through • "*thinks win-win*:" trusts that mutually beneficial outcomes are possible and desirable

Overview of Activities

The chart below identifies the type of tools which are taught by the activities (Activities A through M) found in this supplement. In many cases the activities address more than one type of tool. Following the description of the activities are the Blackline Masters (A to F) that pertain to the Tool Enhancement Activities.

	Background knowledge	Criteria for judgment	Critical thinking vocabulary	Thinking strategies	Habits of mind
Contributing to the group *Taking an active, effective role in small group activities*		• ***Activity A:*** *Imaging success* • ***Activity B:*** *Looks like/ sounds like* • ***Activity C:*** *Matching criteria with technique* • ***Activity G:*** *The secret mission*		• ***Activity B:*** *Looks like/ sounds like* • ***Activity C:*** *Matching criteria with technique* • ***Activity G:*** *The secret mission*	• ***Activity G:*** *The secret mission*
Developing arguments *Generating and framing effective arguments*		• ***Activity B:*** *Looks like/ sounds like* • ***Activity E:*** *Refining arguments and counter-arguments* • ***Activity F:*** *Probes*	• ***Activity E:*** *Refining argu-ments and counter-arguments*	• ***Activity B:*** *Looks like/ sounds like* • ***Activity C:*** *Matching criteria with technique* • ***Activity E:*** *Refining arguments and counter-arguments*	• ***Activity F:*** *Probes* • ***Activity D:*** *Role reversal*
Presenting ideas *Preparing and delivering effective presentations*		• ***Activity B:*** *Looks like/ sounds like* • ***Activity C:*** *Matching criteria with technique* • ***Activity H:*** *Fishbowl* • ***Activity I:*** *Productive peer critique*		• ***Activity B:*** *Looks like/ sounds like* • ***Activity C:*** *Matching criteria with technique* • ***Activity H:*** *Fishbowl* • ***Activity I:*** *Productive peer critique*	• ***Activity H:*** *Fishbowl*
Negotiating agreement *Preparing for and taking part in effective negotiations*		• ***Activity B:*** *Looks like/ sounds like* • ***Activity C:*** *Matching criteria with technique* • ***Activity H:*** *Fishbowl*		• ***Activity B:*** *Looks like/ sounds like* • ***Activity C:*** *Matching criteria with technique* • ***Activity H:*** *Fishbowl* • ***Activity J:*** *Three-tiered bargaining* • ***Activity L:*** *Thinking win-win* • ***Activity M:*** *Negotiating a dress code*	• ***Activity D:*** *Role reversal* • ***Activity H:*** *Fishbowl* • ***Activity J:*** *Three-tiered bargaining* • ***Activity K:*** *The believing game* • ***Activity L:*** *Thinking win-win* • ***Activity M:*** *Negotiat-ing a dress code*

Tool Enhancement Activities

Activity A:
Imagining success

In small groups or as a class, invite students to consider one or more of the following questions and record their suggested list of attributes:

- Think back to the most enjoyable and most successful group that you have been part of. What things did the group do that made it so rewarding for you?
- Think about a person and a time when you were greatly helped to understand or to succeed at something. What did that person do or say? In general, what are the qualities of a good helper?
- What should the ideal group member do?

Activity B:
Looks like/ sounds like

To help students grasp concretely what is involved in employing a particular strategy (or set of strategies) invite them to consider what the strategy would look like and sound like in practice. The following example focusses on interpersonal strategies, but this activity can be used with any of the other sets of strategies for effectively contributing to a group or, for that matter, to many of the strategies involved in the other three group tasks.

Begin by writing on the board, or on an overhead transparency, a title for the strategy (e.g., "effective interaction"). Create two columns. Label one column, "Sounds like," and ask the class, "What would effective interaction sound like? What phrases would you hear if members of your group were relating effectively with one another?" Record students' responses. Label the second column, "Looks like," and ask students how they might implement this strategy in non-verbal ways: "If I were standing on the other side of the classroom and couldn't hear you, how would I know you were interacting with each other effectively?" Record students' responses. Invite students to employ these verbal and non-verbal strategies while completing a small group task that you assign. Debrief after the assignment has been completed, adding to the chart strategies that students suggest. Edit the chart and transfer the information to poster paper for ongoing reference.

A useful variation of this activity, intended to discourage mechanical use of the suggested words and actions, is to add a third dimension — the qualities that these strategies should embody. The point is not, for example, to get students to offer eye contact *per se* (i.e., not merely to stare at one other), but rather to do so in a sincere and appropriate manner. After completing the two columns as described above, invite students to explore the qualities that should characterize use of these strategies. As illustrated on the following page, form an inverted Y to separate looks like and sounds like. Label the newly created space, "Underlying qualities." Ask students, "What qualities would underlie or be reflected in these actions and words?" Record students' responses.

Effective interaction

Sounds like

- "What do you think, Lee?"
- "That's a good idea."
- "Well done, Sally."
- "Can you help me do this?"
- "Does everyone agree?"

Looks like

- facing each other
- making eye contact
- smiling at one another
- nodding with approval
- listening as one person talks

Underlying qualities

- sincere
- friendly
- respectful
- appropriately employed
- not overused

Activity C: *Matching criteria with technique*

List the criteria that would be appropriate for judging success at a particular task in a column. Match the strategies that may help to meet these criteria in a second column. For example, "clarity" is one criterion of an effective presentation. Strategies that may assist students in developing clear presentations include using specific examples, supplementing oral comments with visual aids, limiting the number of points made, clearly sequencing the talk and so on. The chart at the beginning of the Tool Enhancement Supplement outlines a range of criteria and strategies for four group tasks.

Activity D: *Role reversal*

In developing arguments and counter-arguments, it is useful to see things from others' perspectives. Role reversal is a good activity to introduce this to students. Select an issue about which students have some knowledge and, in pairs, assign opposing positions to students. Part way through the discussion, ask students to switch their roles. (This reversal may be repeated at a later time). Instruct students not to offer arguments that they believe are silly, but instead try to offer plausible reasons even if they do not hold the position themselves. The goal is not to reach an agreement but to use the different perspectives to formulate arguments and anticipate counter-arguments. Also, if students develop a habit of considering opposing perspectives, reaching consensus on issues becomes much easier. Debrief by asking students what they learned about developing arguments from the role reversal activity. Encourage students to see the need to be open-minded—modifying initial positions to accommodate the meritorious aspects of an opposing argument.

Activity E: *Refining arguments and counter-arguments*

Students will be better able to develop well-supported justifications if they have a range of questions to consider when trying to generate reasons to support their position or to counter the arguments presented by an opposing side. Arguments consist of descriptive statements (or "fact statements") and evaluative statements (or "value statements"), each having their own considerations for assessment. Using *Fact Statements and Value Statements* (Blackline Master #A) as a guide, introduce students to the vocabulary of argument and counter-argument, and fact and value statement. The Blackline Master explains several questions that students might ask of themselves when analyzing their own arguments or those of others. Suggest that, when asked to prepare a justification for a position, students use these questions to refine and strengthen their basic arguments and counter-arguments.

Activity F:
Probes

To assist students in framing questions that will probe the arguments put forward by other students, introduce the following types of information requests:

- seek clarification
- seek elaboration
- seek evidence
- probe for consistency
- probe for implications
- probe for alternatives.

Using *Probing Questions* (Blackline Master #B) as a guide, review each type of probe and the examples provided. Encourage students to use this Blackline Master as a guide when helping members of their group develop their own arguments or when questioning the arguments put forward by others.

Activity G:
The secret mission

While completing an in-class group task, assign students a supplementary mission to perform. The purposes of assigning the supplementary role are to raise awareness of the dynamics operating in a group and to sensitize students to different ways in which they can contribute to or impede group effectiveness.

Divide the class into groups of between four and six students. Assign and explain a task you want them to undertake (e.g., complete an activity sheet, create a poster). If the task is to be assessed, it should be for practice marks or carry a small numerical weight.

Before sending students on their way announce that there is one additional wrinkle—each is to have a secret mission to fulfill. Distribute one of six role cards found on *Secret Mission Role Cards* (Blackline Master #C) to each person in each group. The individual should keep his or her mission a secret. Stress that students are to act as they would normally with one additional objective: the mission described on the card. Advise students not to be excessive—the goal is to be *effective* but *discreet* in the mission. Students would have successfully fulfilled their mission if in the debriefing session when their role was announced their fellow group members would comment “Yes, of course, NOW I see what you were up to!”

After students have completed the main task, debrief the secret mission:

- ask students individually to write down their hypothesis about each group member’s mission;
- in their groups, let students share their hypotheses about each person’s mission;
- in a class discussion, share students’ reactions—positive and negative—to various members’ actions;
- discuss any general lessons to be learned and suggestions for future group work;
- in a journal or private notes, invite students individually to reflect on the insights they have had about themselves and others through this activity.

Activity H:
Fishbowl

A strategy for sensitizing students to their manner of arguing is to engage students in a "Fishbowl" activity—one in which a small group sits in a circle discussing an issue, outside the circle the rest of class observes and offers feedback.

- Divide the class into three groups: *discussants, process observers* and *content observers*. There should be approximately five *discussants* and ten or more in each of the other two groups.
- Seat students in their groups in the following formation:
  - *discussants* (group D in the diagram) form a small inner circle facing each other;
  - *process observers* (group PO) form a first-ring outer circle so that they can focus on specific *discussants*;
  - *content observers* (group CO) form an outermost circle, watching and listening to the general discussion.

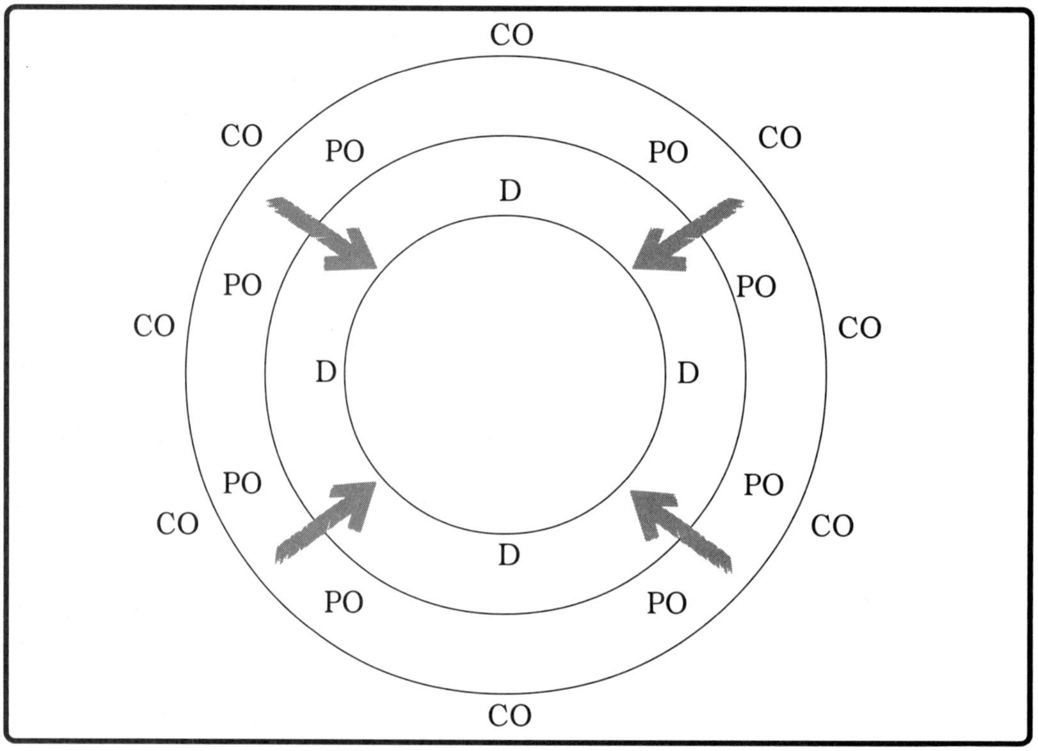

- Assign group tasks as follows:
  - *discussants* engage in a conversation on an assigned issue;
  - *process observers* closely observe and note on a checklist or observation form the discussion strategies (e.g., asks for clarification, makes eye contact with fellow discussants) of the *discussant* closest to them. Approximately two PO's should observe each *discussant*;
  - *content observers* note the content of the discussion—exploring the assumptions, contradictions, arguments, counter-arguments and so on.

After 15 minutes of discussion, invite the *process observers* to report on their observations. *Discussants* may respond to this feedback and consider what they might do to improve their discussion strategies. *Content observers* prepare a report for the class. The report can be given individually, either orally or in writing; alternatively, *content observers* may confer together in preparing a group presentation drawing together the points made into a summary synthesis to present to the class at the end of the discussion.

Activity I: *Productive peer critique*

Students are generally unmotivated to prepare preliminary drafts of written work or to rehearse oral presentations. One way to encourage this is to use the rehearsal or draft as an opportunity for formal *productive critique*—meaning that the critique must lead to improving the final result. The following approach is recommended when introducing students to productive peer critique. It is advisable to model this approach initially in a whole class setting, with a writing sample or presentation volunteered by a student(s) willing to undergo public scrutiny (alternatively, use a sample work or taped presentation by a former student). After the modelling, assign students in small groups to repeat the process with their own efforts.

Review the criteria. Begin by reviewing with students two sets of criteria: first, the criteria related to the *product being critiqued* and then the criteria for a *productive critique*.

- *Criteria for judging the product.* Review with students, perhaps with an overhead transparency or a checklist distributed to each student, a handful of criteria that students are to consider when commenting on a work/presentation (e.g., clear, no extraneous detail, appropriate, worthwhile). Limit the number of criteria to four or five, especially if students are new to the enterprise. You might also ask students to look for specific strategies that contributed or not to satisfying the criteria (strategies for achieving clarity include using specific examples, supplementing oral or written comments with visual aids, limiting the number of points made).
- *Criteria for productive critiques.* Explain to students that just as there are criteria for judging the merits of the work/presentation, so too do productive critiques have criteria for success. Explain the following criteria:
  - *respectful*: comments should be gracious rather than mean-spirited, insulting or condescending (stress the importance of this criterion);
  - *important*: comments—whether positive or negative—should reflect significant strengths and weaknesses rather than trivial, exaggerated or unfounded concerns;
  - *specific*: comments should be clear and explicit, rather than vague and overly general (using an example, illustrate the value of specific as opposed to vague comments);
  - *constructive*: comments should advance the primary purpose of critique which is to improve performance—not to belittle or to criticize; therefore advice on how to improve is preferable to comments that merely note areas of strength and weakness.

Offer the product to be critiqued. While reading the work or listening to the presentation invite students to make written notes in light of the (product) criteria discussed. Initially it may be sufficient to ask students to record their comments in two columns—"Positive Features" and "Areas for Improvement." A more sophisticated approach would include inviting students to subdivide the columns into rows, one for each of the criteria under consideration. Students would then explicitly look for positive features and areas for improvement for each of the criteria.

Set the terms of the critique. Once students have individually reviewed the product, structure the sharing of the critique in the following manner:

- *Lead with the positives.* It is strongly recommended that the first part of a critique consist exclusively of everyone commenting in an *unqualified* manner on the *strengths* of the presentation. The purpose is to reduce the anxiety that all of us feel when we subject our work to the scrutiny of others. Even when reminded about starting with the positives, students have a tendency to slip into qualified (partly negative) comments. To increase the likelihood of a non-traumatic, profitable peer critique, politely interrupt whenever apparently positive comments begin to take on an implied criticism. As well, students will often find it difficult to make positive comments—not because there are no positives in the presentation but because they may be prone to notice negative qualities. It is advisable to prepare beforehand a long list of positive comments to infuse into the discussion in the event that student-generated comments are not forthcoming. Remind students who are offering suggestions that the point is not to show off or to make others feel inadequate. Everything they say and do should be tempered by one guiding purpose: to help make the product the best possible. Only after abundant expression of the strengths, such that the individual is on the verge of being embarrassed with the numerous comments, should concerns or areas of improvement be expressed.

- *Assign a keeper of the standards.* Assign a student (or perform this role oneself) the task of keeping track (perhaps though use of a checklist) of which criteria for the product are mentioned during the critique. For example, students may focus on clarity and appropriateness, but omit worthwhile content. After a period of time, the "keeper of the standards" might draw the group's attention to the criteria that have not been adequately addressed.

Direct the recipient of the critique. Just prior to airing the comments offer the following advice to the recipient(s) of the critique.

- *Active listener.* Encourage the student (or group of students) receiving the advice to assume an active listening role—limiting comments to asking for clarification or elaboration, checking for understanding and whether or not ideas offered by individual commentators are widely held. Encourage recipients not to be defensive nor to defend their work. Their role is to hear what others have to say, and after the critique is over, to decide for themselves which comments, if any, are worth acting upon. Praise them for their courage in volunteering to subject their work to public scrutiny.
- *Critique the critique.* Another strategy for reducing recipient stress is to ask at several points during the critique, including during the sharing of positive comments, for the recipient(s) to critique the critique. In other words, invite the recipient(s) to comment on how well the class is doing in meeting the criteria of a productive critique (i.e., Are the comments respectful, important, specific and constructive?). Not only does this strategy remind the class of the criteria they should use in framing their comments, but it affords the recipient a reciprocal opportunity to comment on the class's performance.

Debrief. Close the activity by asking for student observations, beginning with the recipient(s) of the critique. The teacher's role throughout should be to acknowledge that although peer critique can be difficult, especially for recipients, the experience can also be fruitful and satisfying when the criteria of productive critique are well met. Encourage students to integrate the comments from the critique into their final written written draft or formal oral presentation.

Activity J:
Three-tiered bargaining

Introduce a three-tiered bargaining strategy that includes an opening position, areas open to negotiation and a bottom-line position (no room left to negotiate, time to walk or get help from a mediator). When students are to design a bargaining strategy, provide worksheets that ask them to frame their bargaining strategy in these three stages. Once these tiers have been identified, ask students to justify each strategy (this involves research about other parties' positions, backgrounds, possible negotiating tactics) and generating action plans for each stage of the negotiation. For example, if negotiations have reached a bottom line and nothing more can be traded, the negotiating team needs to list the options open to them: involve a third party (who will remain neutral and mediate), leave and wait to be chased (this move works only if you have something they want), revisit the points of agreement, or reconsider what can be traded and what cannot.

Ask students to make a list of negotiating demands as if they were items to be traded. Once this list is generated, ask students to organize these items in priority sequence. Finally, ask students to identify those items that are non-negotiable and determine whether this is to be kept confidential or revealed while negotiations are in progress.

Activity K:
The believing game

One of the more important elements in reaching consensus is open-mindedness. This habit of mind can be encouraged through Peter Elbow's "Believing Game." Invite or assign a spokesperson (an actual representative or a mere role-player) to present an *unpopular* view to the class, such as joining the United States or advocating nuclear war. Prior to the presentation, ask students to outline or note their (contrary) views on the issue (e.g., make a list of reasons for Canada to remain independent or to eliminate nuclear weapons). During the presentation, ask students *temporarily* to suspend their personal beliefs and actively try to embrace the speaker's unpopular ideas. Students are to silence disagreement inside their heads and try to accept this different perspective. Encourage them to ask supportive questions of the representative that might help them accept this unpopular position. For example, they might want to ask the following questions: "Tell me more about how you came to this position." or "Help me to understand why you believe this?" As the questioning continues, students affirm aloud those parts of the position that they can now accept. For example, they might say, "I can accept that the United States has a powerful and rich culture." Throughout the game students are to silence all internal voices of disbelief. Gradually, the class reaches a consensus as to what parts of the position they can accept.

After the presentation and questioning of the representative, debrief the activity in two parts.

- *Student reactions to the game.* Ask students open-ended questions about their feelings or reactions to having to suspend their beliefs: "How did you feel while listening to the spokesperson?" or "What was difficult for you?" or "What surprised you about this activity?" Most importantly, ask how students' thinking changed. Sensitive debriefing is essential since the exercise may develop considerable anxiety—the more deeply held the position students were asked to suspend, the greater the potential for anxiety.
- *Revisiting their original views.* In the second part of the debriefing, it is also important to invite students to re-examine their original beliefs. Ask students to review their list of reasons and make any changes they feel are warranted. If the game has been successful, there is likely to be some modification of original

beliefs. The goal is not to get students to switch their position, but merely to deepen and render more complex their beliefs.

Activity L: ***Thinking win-win***

Steven Covey, author of *The Seven Habits of Highly Effective People*, talks of "Think Win-Win"—the habit of mind when confonted with a conflict of believing that mutually beneficial results are both possible and desirable. This habit is especially helpful when seeking negotiated solutions.

To introduce a "Think Win-Win" habit of mind invite two students to come to the front of the class to engage each other in a friendly, one-minute arm wrestling match (not sitting at a table, but standing with their arms in mid-air). Announce that during the minute every time one student brings the opposing student's arm to a horizontal position you will award that person a prize (e.g., one minute of free time, five cents). Two other students are to keep a tally of the number of times their candidate scores a take-down. Unknown to the class you have explained beforehand to one of the competitors that he/she is to resist for about ten seconds and then suddenly let his/her arm relax, allowing the other student to win repeatedly and quickly. After ten seconds of letting the other student win, the coached student should gradually try to win him/herself on an alternating basis. The ultimate objective is to create a pattern where both students' arms are fluidly going back and forth from one horizontal position to the other, with each side winning easily in turn. Practice the three-steps (i.e., initially resist, then allow, finally alternate) with the coached student beforehand. If the demonstration is successful it should be obvious to the class that both sides are ahead of the game—even the stronger of the two students is better off when each side lets the other win, since no time is spent struggling with one another.

In teaching students about Thinking Win-Win, explain three assumptions that impede this mindset: (1) an exclusive preoccupation with one's own interests, (2) ignorance or confusion surrounding the other party's position and (3) a zero sum mentality—the belief that anything gained by one party necessarily takes away from the other side.

Offer students the following strategy when confronted with opposing positions for overcoming these impediments:

1. Each side states its position—what it wants (e.g., Québec wants to separate from Canada) and why—its reasons, values and assumptions, or what is often called the interests behind the position (e.g., Québec values linguistic freedom, autonomy to direct external affairs and respect from Canada and other nations). Identifying the interests allows different sides to see the root of issue and to find common ground.
2. Each side carefully and sensitively restates the position and interests of the other side, summarizing these on chart paper as opposing ends of a continuum: the positions are recorded above the continuum at each of the poles; beneath each position are recorded the interests underlying that position.
3. Students then identify other positions that fall between the two poles (e.g., Québec separating but keeping some things in common with Canada, such as money and defense) and place them on the continuum. Beneath each position students identify the interests behind these positions (e.g., avoid duplication of services, direct control of important domestic matters).
4. Together, the sides suggest at least three solutions to the dispute while attempting to maximize the most important interests of each side.

Illustrate this four-step strategy using an everyday classroom incident (e.g., two students wanting to use a computer at the same time) or a familiar current event (e.g., Québec separation).

Once students have understood the principle of Thinking Win-Win and the steps involved in the interests strategy, invite them to apply this approach to a situation in their own lives or from a conflict taken from literature, history or current events. Provide pairs of students with *Plotting Positions* (Blackline Master #D) which contains a continuum and boxes, where positions and underlying interests may be recorded. (The activity can be made easier by supplying various positions on the continuum and inviting students to identify the underlying interests for each position). Share results of several pairs of students with the entire class on an overhead transparency. Discuss proposed solutions and results of adopting the Think Win-Win mentality and the interests strategy. Assess each pair on its ability to successfully complete the activity sheet.

Activity M: *Negotiating a dress code*

To introduce students to a few basic strategies for reaching group consensus engage them in a role play discussion. The following role play uses the issue of negotiating a dress code policy for a school. This activity also encourages students to appreciate the diversity of points of view that exist even within groups that appear outwardly to have similar interests and perspectives.

Begin by asking if anyone has heard of the phrase "reaching consensus." Create a mind-map of student-generated ideas on the board or overhead. Be careful about capturing every one's ideas and about clearly recording their ideas as you are modelling this tool for students. Model for the class another proposed strategy by restating in summary form students' comments and confirming with students that you understand their idea correctly.

Distribute copies of the *Consensus Decision Making* (Blackline Master #E) and review the seven suggested strategies to facilitate group consensus. Indicate that these are some of the simpler techniques; invite students to suggest others.

Explain to students that each of them are to assume one of six roles on a controversial issue. Students are to portray their assigned role's points of view and try to reach consensus on the following issue:

> Imagine that your school is contemplating the implementation of a dress code. You have been asked to sit on an advisory committee that will report to the school board. You have been selected because you represent a particular point of view. Bearing these points of view in mind, you are asked to come to a consensus decision in your group about whether or not the school should implement a dress code. You will be required to elect a spokesperson to report on your decision and the details that support it to the rest of the class.

Assign students into groups of from four to six students and distribute the *Dress Code Role Cards* (Blackline Master #F). Students are to read their role cards individually and not show them to one another. Remind students that they are to use the earlier discussed techniques, (e.g., taking turns, summarize and confirm) to facilitate reaching consensus.

Allow approximately 20 minutes for the role play. As a class, debrief the experience. Ask general questions such as:

- Did any group come to a consensus? What was it? If different groups made different decisions, find out why.
- Did any group not come to a consensus? Why not? What impeded coming to consensus?
- Did anyone in any of the groups feel that his/her views were misunderstood or ignored by others in their group? Discuss what might be done to correct this problem.

Using Blackline Master #E as a guide, invite students to comment specifically on how each of the suggested strategies contributed or not to group consensus building.

This activity can be modified by adopting *Activity G: Fishbowl* as the structure for monitoring the negotiations over the dress code.

Fact statements and value statements

When attempting to persuade someone, typically we will provide arguments or sets of statements to support our position. Arguments are generally made up of a mix of two kinds of statements: fact statements and value statements.

Fact statements

In arguing for a position, persons make statements that they claim are factual. These statements are *descriptive* in that they state what the person believes *is*, *was* or *is likely to be*. For example, the following statements are all descriptive statements:

- Most Albertans support giving Québec special powers.
- Four British North American colonies joined in Confederation in 1867.
- Canada is likely to break up within the next twenty-five years.

We use the terms "a fact" or "factual" to refer to fact statements that are true. The statement, "Most Albertans support giving Québec special powers," does not appear to be factual, since public opinion polls suggest that this is not the view of most Albertans. It is, nevertheless, a descriptive statement, but it is a false descriptive statement since it incorrectly describes the situation. In building arguments, you will want to be sure that the fact statements made are actually supported by the available evidence. More generally, there are several questions you might consider about fact statements to help develop arguments that (1) support your own position and (2) counter the arguments that others put forward in support of their position.

Questions to consider about fact statements

	Developing arguments	Developing counter-arguments
Believable	Are all of my fact statements believable? Can I locate further information to support or cast doubt on my fact statements?	Are all of their fact statements believable? Can I locate information to raise doubts about their fact statements?
Relevant	Which fact statements are relevant to my position?	Are their fact statements relevant to their position?
Contradictory	Do some fact statements contradict each other? How can I explain this?	Do some of their fact statements contradict each other?
Adequate	Do my fact statements strongly justify my position? What else must I prove before my position is justified?	Do their fact statements strongly justify their position? What else must they prove before their position is justified?
Alternative interpretation	Is my interpretation of what follows from the fact statements the most plausible one?	Can I interpret their fact statements differently, in a way which does not support their position?

Value statements

In arguing for a position, persons also make statements about the merits or value of a situation. These statements are either *prescriptive* in that they state what the person believes *should* or *should not be* the case, or they are *evaluative*, in that they state what is *good or bad, right or wrong, excellent or poor,* and so on. For example, the following statements are all value statements:

- Albertans *should* support giving Québec special powers.
- The four British North American colonies *were right* in joining in Confederation in 1867.
- It would be *unfortunate* if Canada broke up.

In building arguments, we want to be sure that the value statements are defensible. Typically we do this by assessing the implications of the value statement in light of other values that we hold. For example, we might assess the defensibility of the value statement, "It would be *unfortunate* if Canada broke up," by imagining all the values that would be affected by Canada's break up. These include basic rights of individuals to direct their own affairs, the economic well-being of the country, the physical and psychological well-being of people. We would then explore the implications for these values, asking ourselves: Would Canada's break up have more serious negative consequences for these values than would its continued existence as a nation (or some other option)? Notice in answering these questions, numerous fact statements about what is likely to occur will arise and these, too, should be examined for their reasonableness. Regarding value statements, the following questions may help you develop arguments that (1) support your own position and (2) counter the arguments that others put forward in support of their position.

Questions to consider about value statements

	Developing arguments	**Developing counter-arguments**
Fair to all	How are my value statements fair to each of the other parties who may be affected by the outcome, especially those who might be harmed?	Are their value statements unfair to any other parties who may be affected by the outcome, especially those who might be harmed?
Consistent with basic values	What other basic values are supported if we accept each of my value statements?	What basic values are ignored or threatened if we accept each of their value statements?
Sets a precedent	How would accepting each of my value statements provide a positive example for settling other situations?	What other negative consequences will follow if their value statements were accepted as setting precedents for other situations?

Probing questions

Use the following chart as a guide to framing respectful questions that will help you and others to understand more clearly the arguments presented and to probe them more deeply.

Purpose of question	Examples
Seeking clarification	• *I'm not clear what you mean by* [a statement or phrase]. *Could you say more?* • *Can you think of an example of what you mean?*
Seeking elaboration	• *Could you say more about this matter?*
Seeking evidence	• *What are some of your reasons for this* [a position]? • *Help me understand why you believe* [a position] *is correct or desirable?*
Probing for consistency	• *You seem to be saying* [a statement of position or evidence]. *How is this consistent with what you said earlier about* [a previous statement]?
Probing for implications	• *If we accept what you say about* [a statement of position or evidence], *what implications would this have for* [another situation or group]?
Probing for alternatives	• *You have argued for* [a solution to a problem or explanations of events]. *Would there be any other alternative* [ways of meeting their requirements or explanations for the situation]?

Tool Enhancement

B

Secret mission role cards

Duplicate and cut sufficient copies of the following roles cards so that each student within each assigned group will have a different mission.

Mr./Ms. Quick N. Dirty

Your mission is to encourage the group to get the assignment over and done with quickly. Do not be compulsive or hostile, but at varying points say things casually to speed up things (e.g., "It's just a class assignment; let's just get it done and have fun").

Mr./Ms. Dug Inn

Your mission is to try, at perhaps two points during the assignment, to get your own way. Stick to your position(s) as long as you feel comfortable doing so. Perhaps come back to it later in an attempt to have things done as you would like them. Don't try to impose silly ideas—you should think them worthwhile.

The Strong Silent Type

Your mission is not to speak unless you are directly asked a question, and even then you should be very brief unless you are encouraged to say more. Be discreet about not saying much and cooperate (silently) in any way you can.

Mr./Ms. N. Courager

Your mission is to acknowledge and encourage fellow students whenever you think their contributions were useful. Do not be insincere and do not overdo it. Think of verbal and non-verbal ways of fulfilling your mission.

Mr./Ms. Consensus Checker

Your mission is to ensure that everyone's opinions have been heard and that everyone is more or less satisfied with the group's decisions on the big issues. Do not be too obvious about your efforts to check for consensus.

The Keeper of the Standards

Your mission is to ensure that all the requirements for the assignment are considered by the group and successfully addressed. Do not be obsessive, but do try to help the group focus on meeting the assigned criteria.

Plotting positions

POLAR POSITION

UNDERLYING INTERESTS

INTERMEDIATE POSITIONS

INTERESTS SERVED

POLAR POSITION

UNDERLYING INTERESTS

Tool Enhancement **D**

Consensus decision making

Consensus means that everyone must agree. A number of strategies may help your group reach a consensus on an issue.

- **Treat others with the respect** that you would expect for yourself. People often may have widely divergent points of view. Insults and other disrespectful comments will further inflame the situation and will not help in coming to an agreement.
- **Listen carefully** to what the other students are saying.
- Make sure you understand other views by **summarizing** what they say in your own words. Ask them to **confirm** that this is what they meant. Many misunderstandings occur because of miscommunication.
- **Take turns** so that everyone gets an equal opportunity to share his or her opinions.
- **Make sure your point of view is heard**. It is not a consensus if your views are not represented and you are uncomfortable with the decision.
- **Look for alternative ways** to meet your goals and those of other people in the group. You may find it helpful to distinguish between **the underlying goals** (what someone is hoping to achieve) and **the means or strategy** (how they propose to achieve their goals). Often, merely by altering the strategy, more underlying goals can be accommodated.
- At some point in the discussion you may find it useful to **make a chart** of the main issues of contention between your points of view. You might also consider trying to **brainstorm** possible solutions to the dilemma by creating a **mind-web**. (A mind web is an illustration or picture of the connections between ideas. Try not to censure ideas even if you do not agree with them. When making a mind-web write down everything everyone says and connect the ideas with lines. Afterwards eliminate those ideas that are unacceptable to some members.)

Dress code role cards

Duplicate and cut out sufficient copies of the following role cards so that each student within each assigned group has a different point of view.

Gothic

You are a grade 11 student. You really enjoy listening to Marilyn Manson's music and want to wear clothing that reflects your taste in music. You often wear Marilyn Manson t-shirts. You cannot see anything wrong with these t-shirts, and you think it is ridiculous for people to be upset by them. Even if you could afford to buy expensive clothes you would not. Your black clothing, dark makeup and Gothic jewelry are an expression of your personality—your identity. You do not think anyone should have the right to tell you what you can and cannot wear. Besides, uniforms are unattractive. They make everyone look the same when in reality everyone is unique. You take pride in being different and think that people should be allowed to express their uniqueness through their choice of clothing.

Designer clothes

You are a grade 11 student. You wear only the best designer clothing. You can buy whatever you want whenever you want to. You and your family have earned the right to wear expensive clothing. Your mother and father have worked hard to have the social position that they do and they should be allowed to show it. You think that it is good idea to wear expensive clothing because everyone is constantly making impressions on people. You can tell if someone is refined and has good taste by the clothes they wear. Who knows what opportunities will open up to you because of a favourable impression you make? All of your friends wear the same kind of clothes. Uniforms would make you look ugly. It would be wrong for the school authorities to force you to look like everyone else.

Hard times

You are a grade 11 student. Your family is experiencing economic difficulties. Your father is unemployed. Your mother has just gone back to work as a waitress. Money is very hard to get, and you have three younger brothers who need clothes too. Although you have a part-time job at the local corner store, the money you make must contribute to buying food and paying rent for the family apartment. You believe standard uniforms would ease the pressure on you to wear the right clothes. You want to fit in and be respected as a person, but lately you have not been able to keep up with the in-crowd. You would really like to wear expensive designer clothes but you do not think you should be penalized socially at school because your family cannot afford them. You just want to have friends and get through school without too much stress. Uniforms would take some unnecessary pressure off you and your family.

Keep it simple

You are a grade 11 student. You hate getting dressed in the morning. Every day it is difficult for you to decide what to wear because you want to impress people, and you want to look good. You believe that it is very important to look attractive, but lately you have been thinking that it would be a lot easier to get to school on time if you did not have to make so many choices in the morning. A standard uniform might reduce unnecessary competition between people on the basis of what they wear. People could still express their individuality through different hair styles. And it might get people to looking more deeply at each other—learning to see them for who they are instead of what they wear.

Let kids be

You are a parent of a grade 11 student. You think uniforms are unnecessary and outdated. You went to a private school as a child and had to wear a uniform. The time when people were required to look and dress according to an established standard is long gone. People should be allowed to express themselves any way they want and not be limited by a required uniform. People have to learn how to dress themselves—it is part of their education. You think the idea that girls should wear skirts while boys wear pants is sexist and degrading to women. People should be comfortable with what they wear and have choices. Who would pay for the uniforms anyway? The government? You think tax dollars could be better spent on more important matters.

Let's be sensible

You are a parent of a grade 11 student. You think uniforms would solve some of the problems at your son's high school. There is too much focus on the status associated with clothing at school. Your son is always asking you to buy him new clothes, and they are always expensive. Parents should pay a one-time fee for a school uniform that would last for the whole of the student's career. The uniforms could be recycled from year-to-year as students grow out of them. Students who ruined a uniform would have to pay for a replacement. In addition, you think there should be rules about the length of hair for boys and the style of hair for girls. Girls should wear plaid skirts and white blouses that button up to the neck. Boys should wear long blue pants, white shirts and plaid ties. You believe this would be a great equalizing force in the education system and place the focus where it belongs: on studying and learning, instead of on appearances.